Springer Praxis Books

Popular Astronomy

This book series presents the whole spectrum of Earth Sciences, Astronautics and Space Exploration. Practitioners will find exact science and complex engineering solutions explained scientifically correct but easy to understand. Various subseries help to differentiate between the scientific areas of Springer Praxis books and to make selected professional information accessible for you. The Springer Praxis Popular Astronomy series welcomes anybody with a passion for the night sky. Requiring no formal background in the sciences and including very little to no mathematics, these enriching reads will appeal to general readers and seasoned astronomy enthusiasts alike.

Many of the books in this series are well illustrated, with lavish figures, photographs, and maps. They are written in a highly accessible and engaging style that readers of popular science magazines can easily grasp, breaking down denser aspects of astronomy and its related fields to a digestible level.

From ancient cosmology to the latest astronomical discoveries, these books will enlighten, educate, and expand your interests far beyond the telescope.

Matthew McMahon • Pedro M. P. Raposo
Mike Smail • Katherine Boyce-Jacino
Editors

100 Years of Planetaria

100 Stories of People, Places, and Devices

Editors
Matthew McMahon
Armagh Observatory
College Hill
Armagh, Ireland

Pedro M. P. Raposo
The Academy of Natural Sciences
of Drexel University
Philadelphia, PA, USA

Mike Smail
Adler Planetarium
Chicago, IL, USA

Katherine Boyce-Jacino
Adler Planetarium
Chicago, IL, USA

Springer Praxis Books
ISSN 2626-8760   ISSN 2626-8779   (electronic)
Popular Astronomy
ISBN 978-3-031-75495-1   ISBN 978-3-031-75496-8   (eBook)
https://doi.org/10.1007/978-3-031-75496-8

Adler's first Director, Philip Fox, operates the ZEISS model II for an audience, 1930-1937 (Kaufmann & Fabry, Adler Planetarium)

This Springer imprint is published by the registered company Springer Nature Switzerland AG
The registered company address is: Gewerbestrasse 11, 6330 Cham, Switzerland

If disposing of this product, please recycle the paper.

Adler's first Director, Philip Fox, operates the ZEISS model II for an audience, 1930–1937 (Kaufmann & Fabry, Adler Planetarium)

Preface

The planetarium as we know it today is a relatively recent invention and forms part of a very long story of humans attempting to build immersive spaces and to popularize the science of astronomy. A planetarium projector is a device, or instrument, that, by means of optical projection, turns the inside of a large domed space into an illusionary sky. The projector itself may be hidden in the walls or it may, as it did over one hundred years ago, sit in the center of the space. Physically in this case, as well as metaphorically, the projector is the beating heart of any planetarium. The dome, the canvas (sometimes literally) onto which it projects the stars and planets, is an expansive space in which the visitors are invited to step into a fully controlled night sky. The didactic purpose is clear, to provide a sky that is perfectly clear, and can be moved forward and backward in time, at the control of a lecturer who explains the constellations and motions to an audience.

In the last century, the dome has grown and matured within that didactic purpose, with new ways to engage the visitor and a very different understanding of our own place in the universe. Stories can now be told with the dome that are dynamic and reactive, and the audience are increasingly part of that story itself. Outside of the planetarium's purpose as a teaching aid, the dome has been a site of artistic expression and contemplation for the public. Planetaria have grown alongside observatories and science centers to help their local communities and engage successive generations. New directors and staff have tackled increasingly complex themes and continue to develop immersive ways for the visitor to interact with the dome and the projector.

In 2023, we began the celebrations of the first centenary of the projection planetarium, built in Jena, Germany, in 1923. The centenary has been a celebration of the entire sector, but also a chance for individual institutions to reflect on their own story. For some institutions, it has been a story of great successes, pioneering new technology and inspiring generations of new astronomers. For some, it has been a story of struggle against great odds, driven by a passion to show the public what has been hidden from their view by clouds, light pollution, and a changing landscape. The buildings that house these instruments of the illusionary sky have changed dramatically, and many reflect their long histories in organic and sprawling additions. Other planetarium projectors live their lives on the road, transported all over their native countries to show the heavens in small portable domes.

The people that have cared for these machines, and used them, and continue to work in planetaria around the world, are as varied as the institutions themselves. They come from all walks of life, some are trained astronomers, some are not. They are a wildly diverse sector but drawn together by their intimate knowledge of their machines and the stars they portray. Their stories are collected and cared for by curators and historians who have seen the value in their diversity of experience and will preserve them for future generations. As the sector faces another centenary, it may find solace in those that came before and a sense of community that transcends time and place.

This volume is a collection of stories that stretch across the pre-history of planetaria, to visions of the future. We have included biographies of the people, places, experiences, and objects which come together in planetaria across the world to create unique and beautiful interconnected networks. It would not have been possible without the many individuals who contributed their stories to this work and those who supported it with their ideas and assistance.

Armagh, Ireland   Matthew McMahon
Philadelphia, PA, USA   Pedro M. P. Raposo
Chicago, IL, USA   Mike Smail
Chicago, IL, USA   Katherine Boyce-Jacino

Contents

Objects of the Planetarium   1
Liba Taub, Joshua Nall, Michael Burton, Heather Alexander,
Felix Luhning, Pedro M. P. Raposo, Katie Boyce-Jacino, Kurt Vanhoutte,
Mike Smail, Gary Likert, Helen Ahner, David C. Leake, Lionel Ruiz,
Michael McConville, Garry F. Beckstrom, Bing Quock,
Stefano Giovanardi, Cora Braun, Terence Murtagh,
Volkmar Schorcht, and Matthew McMahon

Building the Planetarium   67
Yann Rocher, Katie Boyce-Jacino, E. C. Krupp, Pedro M. P. Raposo,
Mike Smail, Cora Braun, Misa Ichikawa, Paul McFarlane,
Terence Murtagh, M. A. Rosario C. Ramos, Yuliia Prybytkova,
Owen Phairis, Daniel-Chu Owen, Suzi Murabana, Tomáš Gráf,
Ondřej Smékal, and Matthew McMahon

Cultures of the Planetarium   105
Anna Gammon-Ross, Matthew McMahon, Pedro M. P. Raposo,
Terence Murtagh, Bing Quock, Mike Smail, Stefano Giovanardi,
Jane Kanter, Charlotte Bigg, Susanne M. Hoffmann, Paul Cornish,
Helen Ahner, Katie Boyce-Jacino, Michael G. Neece, David DeVorkin,
and Andreas Schmidt

People in the Planetarium   145
Ben Buhl, Yaël Nazé, Aubrey Henrietty, Michelle Nichols, Volkar Schorcht,
Pedro M. P. Raposo, David DeVorkin, Andrew Johnston,
Katie Boyce-Jacino, Mike Smail, Nigel Henbest, Stephanie Ridley,
Noreen Grice, Hachioji, Wolfgang Steffen, Nico Koning,
Kerem Osman Çubuk, Stefano Giovanardi, Chris Helms, Arjun Chawla,
and Ka Chun Yu

Objects of the Planetarium

Liba Taub, Joshua Nall, Michael Burton,
Heather Alexander, Felix Luhning, Pedro M. P. Raposo,
Katie Boyce-Jacino, Kurt Vanhoutte, Mike Smail,
Gary Likert, Helen Ahner, David C. Leake, Lionel Ruiz,
Michael McConville, Garry F. Beckstrom, Bing Quock,
Stefano Giovanardi, Cora Braun, Terence Murtagh,
Volkmar Schorcht, and Matthew McMahon

1   Introduction

Planetaria are sites of material culture that act as an intersection between technological instrumentation, media and astronomy. The various instruments, from the pre-history of the planetarium, before the optical-mechanical projector was invented, tell us how generations learned about science. Many of these instruments are handmade, some demonstrate just how far and fast

L. Taub • J. Nall
University of Cambridge, Cambridge, UK

M. Burton • H. Alexander • T. Murtagh • M. McMahon (✉)
Armagh Observatory and Planetarium, Armagh, Northern Ireland, UK
e-mail: matthew.mcmahon@Armagh.ac.uk

F. Luhning
Zeiss-Großplanetarium, Berlin, Germany

P. M. P. Raposo
Academy of Natural Sciences of Drexel University, Philadelphia, PA, USA

K. Boyce-Jacino • M. Smail
Adler Planetarium, Chicago, IL, USA

K. Vanhoutte
University of Antwerp, Antwerp, Belgium

1

technology has changed in one hundred years. Some may be very familiar to those planetarians (those work in planetaria) who have taken an interest in the history of their sector, and some have never been documented before.

The planetarium as a conceptual space is centered on the projector system, both physically (until very recently when projectors for digital systems were moved into the walls) and metaphorically. As such the aesthetic, technology and design languages of these projectors reflect the changing priority of their eras, their makers and their users. Crucially a planetarium projector is a vehicle for modification, and their users are quick to modify it to reflect their needs for scientific accuracy, or spectacle.

But a planetarium is also more than the leviathan in the middle of the dome, it is also the evolution of a long history of armillary spheres, orreries and magic lantern projectors. Moreover, the planetarium is built on a whole ecosystem of technologies outside of the dome that are part of the visitors' experience. This chapter aims to highlight these objects and tell their stories.

G. Likert
Home Planetarium Association, Gallatin, TN, USA

H. Ahner
Berlin-Brandenburg Academy of Sciences and Humanities, Berlin, Germany

D. C. Leake
William M. Staerkel Planetarium, Parkland College (retired), Champaign, IL, USA

L. Ruiz
Marseille, France

M. McConville
COSM, Chadds Ford, PA, USA

G. F. Beckstrom
Delta College Planetarium, Bay City, MI, USA

B. Quock
California Academy of Sciences, San Francisco, CA, USA

S. Giovanardi
Planetarium and Astronomical Museum of Rome, Rome, Italy

C. Braun
Kiel University, Kiel, Germany

V. Schorcht
Carl Zeiss Jena Planetarium Division (retired), Jena, Germany

2   Ancient Astronomical Globes

Liba Taub

The ancient Greek word *sphairopoiïa* meant 'sphere-making', describing the construction of astronomical globes as well as objects designed to model the motions of the Sun, Moon and wandering stars, the planets. Classicists have suggested that Plato's fourth century BCE Academy may have possessed a model of the cosmos, based on statements in his dialogue *Timaeus*. Timaeus claimed that it is impossible to describe the motions of the planets without visible models; the instrument referred to may have corresponded to some type of armillary sphere. Armillary spheres, composed of rings (from Latin *armillae*) representing the great circles of the celestial sphere, were apparently used in antiquity both as teaching or demonstration models and observing tools.

Marcus Tullius Cicero (106–43 BCE), the Roman politician and orator, referred in several works to ancient Greek astronomical models. His dialogue *The Republic* shares information about several objects. One of the interlocutors, Philus, described a celestial globe credited to Archimedes of Syracuse (ca. 287—ca. 212 BCE), which he had seen in the home of his colleague Marcus Marcellus, whose grandfather had carried it away after the defeat of Syracuse in the Second Punic War, in 212 BCE. He mentioned that an even more impressive and well-known celestial globe, also made by Archimedes, had been installed in the temple of Virtue by the elder Marcellus.

A brief history of astronomical globes is credited by Philus to Gallus, a friend of Cicero, who described solid celestial globes constructed by Thales of Miletus (flourished 586 BCE) and Eudoxus of Cnidus (408–355 BCE). These are contrasted to a 'newer' kind of globe depicting the motions of the Sun, Moon and other planets. Archimedes' invention was even more noteworthy, as he represented the various movements and their different rates of speed with a single device that could be turned. Philus reports that when Gallus set in motion this bronze device, the Moon was always as many revolutions behind the Sun as the number of days it was behind it in the sky. Furthermore, solar eclipses were also depicted.

Cicero also briefly mentions a planetary model by Posidonius (ca. 135–51 BCE) in *On the Nature of the Gods*. Writing sometime later, the poet Ovid (43 BCE—17/18 CE), in the *Fasti*, describes a Syracusan astronomical globe as a temple ornament, a powerful symbol of the cosmic order. Although the accounts of some of the astronomical models may be apochryphal (especially that credited to Thales), the references in Cicero's works to celestial globes and planetaria suggest wider awareness of and interest in such devices.

With the exception of the so-called Antikythera Mechanism (the fragments of which are in the National Archaeological Museum in Athens), no ancient Greek device modelling astronomical motions survives. The Farnese Atlas, a marble sculpture of the god Atlas holding a celestial globe illustrating constellations, housed in the National Archaeological Museum of Naples, is a Roman copy of an earlier Greek example. The Farnese Atlas does not show individual stars and the celestial circles are inexact, suggesting that the globe itself was meant to be evocative of astronomical and cosmological ideas, rather than a scientific tool.

3   Roger Long's 'Uranium'

Joshua Nall and Liba Taub

Roger Long's 'Uranium' counts as one of the more unusual devices to have ever been constructed at the University of Cambridge. A celestial sphere measuring eighteen feet in diameter and weighing over one thousand pounds, Long's invention was built in a room at Pembroke College in or around 1758. Made of iron ribs with tin plating, the sphere was entered via steps at the south pole and was capable of seating thirty occupants. The inner surface was painted to represent the stars and constellations and the entire sphere could be turned by a winch.[1]

Long designed the sphere as an interactive space for lecturing on the general principles of astronomy. Elected as the first Lowndean Professor of Astronomy and Geometry in 1750, Long carried an unusually large teaching load by the standards of the day: at least forty lectures a year split across the two subjects.[2] It seems likely that he was chosen for the post in part because of his aptitude as a teacher and public speaker—a skill he had developed through his very active work as a priest and rector.

Certainly, Long exerted a great deal more effort popularising his subject than he did making observations. His Astronomy, in Five Books (1742–84) aspired to "make the knowledge of the heavenly bodies as easy as the nature of the thing will admit." For this task Long favoured demonstrational devices,

---

[1] Taub (2004), p. 175–77.
[2] Clark (1904), p 202–6.

several of which are illustrated and described in his textbook. The frontispiece of the first volume shows his 'glass sphere', a tabletop globe engraved with constellations and enclosing a smaller globe of the earth. Both the earth and celestial sphere could be rotated independently, a design that Long claimed was original to him.[3]

For lecturing to larger groups Long favoured much bigger models. His first construction at Pembroke was a twenty-foot diameter metal hoop showing the constellations of the zodiac around the ecliptic. Long valued the opportunity to place his students in the position of the observer. "Sitting in the midst" of this hoop, he wrote, "I have the same view of things as if I was looking at so much of the starry heaven itself".[4] His 'Uranium' was evidently the next step in this pursuit of an encompassing celestial sphere viewed from within. In 1742 Long could only state his hope that such a device might one day be constructed; by the second volume of his book, published in 1764, he could triumphantly describe his realisation of this "Great Sphere" at Pembroke.[5]

Long notes that his sphere includes space for "a planetarium" at its centre, though it is not clear whether such a model of planetary motions was ever included in the finished Pembroke sphere. An orrery is illustrated at the centre of the instrument shown in the frontispiece to volume two (Fig. 1)—but this appears to be a much smaller version of the 'Uranium', seemingly tabletop sized, and thus plausibly the version that was advertised for sale by the instrument-maker George Adams Snr. in 1746.[6]

An endowment from Long ensured that the Pembroke sphere continued in use for more than a century. In its latter years it appears to have undergone significant modifications, including "the stars being pierced through the metal according to the several magnitudes".[7] But interest waned and renovations to Pembroke's grounds in the early 1870s necessitated the sphere's removal. It was offered to the nascent Science Museum in South Kensington but refused; a photograph survives of the 'Uranium' taken shortly before its demolition (Fig. 2).

---

[3] Long (1742a), p ix–x; p 71.
[4] Long (1742a), p x.
[5] Long 1742b), p iii–iv.
[6] Adams (1746), p 243–63.
[7] Smyth (1844), p 179.

Fig. 1 A tabletop version of Long's 'Uranium', from the frontispiece to vol. 2 of his *Astronomy, in Five Books* (1764). Image © Whipple Library, University of Cambridge (STORE 38:7)

Fig. 2 Long's 'Uranium' in the grounds of Pembroke College, 1871, shortly before its demolition. By permission of the Master and Fellows of Pembroke College, Cambridge

4   Gilkerson Orrery

Michael Burton and Heather Alexander

The term 'planetarium' was first used to describe what we know today as an orrery, a mechanical device that provides a model of the Solar System and can be used to depict the relative orbital motions and distances of the planets about the Sun.

The name Orrery was derived from one presented to an English nobleman, the 4th Earl of Orrery, Charles Boyle. Boyle was a Fellow of the Royal Society and the device that now bears his name was made by the clock maker George Graham in 1713 whom Boyle had under his patronage.

By 1766 the Orrery had become a popular tool for education, with an Orrery taking centre stage in Joseph Wright's painting A Philosopher Lecturing on the Orrery.

"Earl of Orrery" is a title in the Peerage of Ireland. This Irish connection links us to the orrery pictured here, the Gilkerson Orrery, a part of the historic instruments collection of the Armagh Observatory and Planetarium in Ireland.

The Observatory in Armagh was founded as an "Observatory and Museum" through a 1791 Act of Parliament. The museum existed to educate guests and visitors with an array of devices, machines and rare books dedicated primarily to science.

James Gilkerson traded at 8 Postern Road, Tower Hill at this time. He was well known in London scientific circles, and would make many telescopes, slide rules and this orrery in conjunction with William Gilbert. This Gilkerson Orrery was purchased in 1819 by Somerset Lowry-Corry, the Second Earl of Belmore of Castle Coole in Ireland. It remained in the family's collection until 1956 when it was donated to Armagh Observatory by Galbraith Armar Lowry-Corry, the Seventh Earl of Belmore.

Of particular note is that the orrery only depicts the planets up to Uranus, as Neptune had not been discovered at that time, i.e. the model reflects the knowledge of astronomy of its period, including the relatively few moons of the outer planets then known.

When Eric Mervyn Lindsay became the Director of Armagh Observatory in 1937, he came with a passionate vision to educate. One of his first projects was to transform the Observatory's Meridian Room (with transit telescopes once used for measuring the positions of the stars) into a museum of astronomy. By the close of the Second World War Lindsay began to plan a much larger project, a planetarium. The Gilkerson Orrery formed an important part of the case Lindsay was building that lead to the construction of Armagh Planetarium in 1968. Photographs of Lindsay surrounded by enthralled school children learning through visual demonstration foretold the work done today at Armagh Planetarium by the dedicated Education team.

Armagh Planetarium is today the longest running planetarium in both the UK and Ireland. Together with the Observatory—also the longest continuously operating astronomical observatory in the British Isles—AOP is known internationally for its contributions to research, education and outreach across four centuries. The orrery played an important role in stimulating this achievement (Fig. 3).

Fig. 3 The Gilkerson Orrery, London, 1814 (Armagh Observatory and Planetarium)

5 The Gottorfer Giant Globe

Felix Luhning

Between 1650 and 1657 an astronomical-mechanical marvel was built. Duke Friedrich III of Schleswig-Holstein-Gottorf (1597–1659) was interested in natural sciences and desired an object that could show the night sky on a large scale: the Gottorf Giant Globe. This marvel quickly became famous far beyond the borders of the country, and was the first walk-in planetarium in world history.

The project was directed by the court scholar Adam Olearius (1599–1671), who was at the same time the prince's educator and a confidant of his master. The development of the product belonged to the congenial Limburg gunsmith Andreas Bösch (life data unknown).

The idea for the globe came from the utopian story Chymische Hochzeit: Christiani Rosencrcutz Anno 1459 by the theologian Johann Valentin Andreae (1586–1654), published in 1616. Using a fictitious first-person narrator to tell the story, Andreae describes a seven-day alchemical initiation rite. In the story, a Giant Globe of about 7.5 m diameter plays a role. The Giant Globe stands, almost half embedded in the floor, in the middle of a wide hall, while two men hidden underneath control the motions of the globe. Countries and places of interest on the earth are marked on the outside of the globe. Observers can enter through a hatch decorated with dedication inscriptions that is located in the ocean. Once inside, the visitor can sit on a round bench and admire the starry sky.

The outside of the 3.1-meter globe depicted earth, while the inside represented the vault of heaven. The viewer could climb in through a hatch and sit on a round bench around a table. The wrought-iron construction for this is attached to the fixed globe axis. By means of a hand crank and gears, the globe could be set in motion and the course of the stars demonstrated, while a mechanical sun also wandered through the zodiac. The constellations were illuminated in color and the stars were represented by silver-gilded, radially filed nail heads that sparkled when a candle was placed on the table.

A special feature was that the globe, like Andreae's, had a second, hidden drive. This second drive was powered by a water wheel and a heavy worm gear reducer instead of human strength.

The globe itself was built as a wrought-iron skeleton construction, covered on the outside with copper sheet and on the inside with wood. Both the exterior and interior surfaces were then covered with canvas primer, on which the cartography was applied. Maps and globes from the Amsterdam publishing house of JOAN Joan Blaeu were used as models for the cartography of the earth and sky. On the outside of the globe there was a wide horizon ring, alternately supported by halls and caryatids, which was accessible via a ladder staircase.

The globe was placed in a large, purpose-built summer residence in the 'Newes Werck' gardens behind the palace. The upper floor of the vacation house served as a summer residence for the ducal family. The ground and main floors were reserved for the globe alone. The basement housed not only the kitchen but also the waterpower mechanism for the globe. Interested people could visit the marvel at any time.

The flat roof was accessible—it served as an overlook, 'party-location,' and also certainly for contemplative astronomical observations. Friedrich III did not live to see the completion of his globe. His successors shared other interests, so that the marvel was soon neglected. In the course of the Great Northern War, the globe was taken as Russian spoils of war to St. Petersburg in 1713, where it can still be seen today—in a greatly altered form—in the former Tsarist Art Chamber. The vacation house subsequently fell into disrepair and was demolished in 1768.

In 2002, a new building of the Globe House was constructed in Schleswig, which also houses a reconstruction of the giant globe. The historical conditions are here admittedly only very difficult to recognize.

6   The Portable Orreries of William Jones

Pedro M. P. Raposo

Before the word "planetarium" became widely used to name the kind of setting with a projection system and a dome celebrated in this volume, it was used to refer to a kind of device that represented and simulated the order of the planets and their motions around the Sun, also known as "orrery." The orrery emerged in the seventeenth century and went through significant developments in the 1700s in the hands of skilful instrument makers such as James Ferguson, George Adams, and Benjamin Martin. In its simplest forms, the orrery shows the Sun-Earth-Moon system; in this case, the device is called a tellurian. The most sophisticated models, known as grand orreries, were meant to provide a fuller picture of the solar system including all known planets, and often all known satellites as well.

Orreries were sophisticated and expensive machines, but by the turn of the nineteenth century, with a growing market for educational devices sustained by a rising middle class, some instrument makers sought to produce affordable machines. Such was the case of William Jones of the London-based firm W. & S. Jones, who designed a portable tellurian that he presented in a booklet published in 1782 titled *The Description and Use of a New Portable Orrery.* The "new portable orrery" was actually based on designs previously marketed by Ferguson and Adams consisting of a hand-driven tellurian with an earth-moon ram carrying wheelwork. The tellurian was sold with a baseboard to which a printed overlay containing a zodiac calendar was affixed. The whole apparatus could be disassembled and fit inside a carrying box.

Jones also produced a larger and more elaborate version of the orrery, in two different sizes, in which the baseboard was raised on legs and a mechanism for the motion of the superior planets could be added. Jones went on to advertise orreries of several sizes, including an all-brass model, and to remark that the firm W. & S. Jones accepted commissions for large, customized orreries "to any price." But the hallmark of Jones's production was his "new portable orrery." This is attested by the successive editions of the aforementioned booklet, which amounted to six, the last dating to 1812. After Uranus was

Fig. 4   Orrery by W. & S. Jones (Adler Planetarium collections W-83)

identified as a new planet by William Herschel in 1782, Jones updated his orrery to account for the newly found orb, which appears in ensuing editions of his booklet and versions of the instrument's baseboard with the name Georgium Sidus, chosen by Herschel (but then dismissed by the astronomical community) to celebrate his patron King George III.

Originally meant to be affordable portable planetariums that would sell in numbers among a middle class willing to invest in educational aids, Jones's orreries are now valuable collectibles that realize high prices in the antique scientific instrument market. They testify to historical efforts not only for making astronomy more accessible to broader audiences, but also for profiting on the allure of astronomy and the sensational character of new planetary discoveries (Fig. 4).[8]

7   The Eidouranian Orrery

Katie Boyce-Jacino

In considering the progenitors of the planetarium, we often gesture to immersive spaces like the Gottorp Globe or panoramas, or to mechanical models

---

[8] Jones (1782).

such as orreries, or public scientific lectures. One little known precursor combines all three of these qualities: a remarkable but little-known device called the Eidouranion Orrery, which was shown to great acclaim on the stage of the Lyceum Theater in London, for about fifty years around the turn of the nineteenth century. The Eidouranion was the invention of the instrument maker Adam Walker and his sons, and unlike its table-top contemporaries, the Eidouranion Orrery was composed of a series of moving slides projected onto a screen which faced the audience and extended from the floor of the stage up to the upper curtains.[9]

The slides included elaborately illustrated images of the zodiac as well as detailed sketches of orbital paths of planets, comets, and moons. The final scene, added sometime before 1820, was of "The Probable Construction of the Universe," in which "The Sublime and awful Simplicity of Nature is daringly imitated."[10] This final scene differed from all the others by radically altering the position of the viewer. For the preceding scenes, the audience views the heavenly mechanisms as if from above, but in the final scene, the viewer is suddenly brought back to Earth, presented not with a view from above but a view from underneath — how the sky would look if all heavenly bodies were visible from earth. The milky way stretches across the canvas, "powdered with stars," bearing an uncanny "resemblance to nature."[11]

The Eidouranion Orrery is remarkable for its combination of scientific attention (the Walker family corresponded regularly with William Herschel to confirm their claims about nebulae and planetary motion) and its theatricality. The vertical orrery was a design not seen before or since, and by all accounts was one of the most popular Lyceum offerings.[12] The Orrery shared the gilded stage of the Lyceum with magic shows, animal tricks, illusions of all types. In the two extant depictions of the Orrery, the theater is crowded and chaotic, with spectators surging forward in their seats to get a better view. The pedantic and entertaining qualities of the apparatus were intentionally balanced for such an audience.

The connection between the Eidouranion and the planetarium, which appeared more than a century later, is made clear in the way each apparatus presents itself. On the frontispiece of the surviving pamphlet on the Eidouranion is an epigraph that claims "Stars teach as well as shine!"[13] We

---

[9] Walker (1795).

[10] Walker (1795).

[11] Walker (1795).

[12] During (2005), p 219.

[13] During (2005), p 219.

find similar descriptions of the Zeiss Planetarium more than a century later. Max Wolf, one of its original designers, reflected in 1927 that the planetarium "has grown to be a popular means of education almost without parallel in any branch of learning within the history of man; a means of education, moreover, which does not dishearten but which fascinates by the enjoyment it provides."[14]

More succinctly, Albert Ingalls, an American amateur astronomer who was one of the first international visitors to the Zeiss Planetarium, wrote that "the Planetarium is a good show. [...] Intrinsically the performance is aesthetic. It provides thrills while it educates."[15]

8   Nineteenth-Century Pocket Planetarium

Pedro M. P. Raposo

Imagine carrying a planetarium in your pocket. Nowadays, that might sound relatively trivial, as mobile phones have become an integral part of daily life and there is a wide array of astronomy and sky apps available to those willing to find their way among the stars without having to sit underneath a dome or peruse a printed star atlas.

Now let's travel in time to the first decades of the nineteenth century, way before scrolling down one's social media feed became a repeated daily ritual. While the digital technologies of today were still beyond the wildest dreams of even the boldest visionaries, the wish to carry the universe around in one's pocket was already being catered for by globemakers. Terrestrial and celestial globes were essential tools in creating the modern image of our planet and in mapping the skies respectively, but they were often expensive instruments meant to sit at the study table of cartographers and astronomers, and more often than not to decorate the salons of wealthy sovereigns and dignitaries. Especially in the latter case, globes were usually presented as a pair, comprising a celestial and a terrestrial globe. Thus the whole universe could be brought inside a room. What about having it all in one's pocket?

In the 1670s, the British cartographer Joseph Moxon came up with the solution, introducing the pocket globe. It comprised a small terrestrial globe inside a round case with two hinged halves showing the northern and southern constellations inside. In the eighteenth century the pocket globe became

---

[14] ASTRO 907, Carl Zeiss Archives.
[15] Ingalls (Sept 1929).

a fixture among traveling lecturers, students, and generally young men willing to show that they cared about science. They were sure to impress those unfamiliar with the device when in social occasions they produced a pocket globe and opened up its round case revealing a small earth surrounded by the two celestial hemispheres.

With the rise of the middle class during the Industrial Revolution, the market for scientific instruments and toys expanded significantly. One of the most spectacular and popular educational instruments in astronomy was the orrery, a mechanical model showing the motions of the planets around the Sun, or in its simplest form, known as a tellurian, the Sun-Earth-Moon system. What if the globe and the orrery could be combined into a single portable piece?

The idea was pursued by Richard Ebsworth, a British maker of scientific instruments, educational models, and toys. The piece shown in the image is a pocket tellurian by Ebsworth dating from 1825 in the collections of the Adler Planetarium. Lifting the round portion of the case reveals a small geared model of the Earth-Sun-Moon system, which could be operated by means of a small handle. The tellurian proper sits on a round baseboard presenting the constellations of the Zodiac. The interior of the round lid shows the northern constellations, a design directly derived from the pocket globe that to a contemporary eye also invokes the image of the night sky projected on a planetarium dome.

The whole assemblage is so small (it can be held on the palm of one's hand) that it is difficult to imagine the device being used as an actual teaching or study aid. Similar to pocket globes, it was likely more of a conversation piece that the proud owner would produce out of their pocket to the amazement of their interlocutors. As a former curator at the Adler Planetarium who demonstrated the piece before contemporary audiences several times, I can attest that almost 200 years on it still causes an impression, never failing to "wow" onlookers.

However, the fact that only a very few pieces of this kind are known to survive suggests that, despite their appeal, they were not particularly popular or commercially successful. The growing availability of affordable and easier to handle didactic materials for teaching and learning astronomy during the nineteenth century probably contributed to that. It is possible that only a few were made and marketed as novelty showpieces. Nevertheless, pocket orreries like the one in the Adler's collections testify to an enduring fascination with maps, globes, and tellurians, and to the wish of having the stars at our fingertips—to which the mobile devices of today can cater much more efficiently, even if with less glamor (Fig. 5).

Fig. 5 Pocket tellurian by Richard Ebsworth, London, England 1825 (Adler Planetarium collections W-51)

9   Magic Lantern Performance

Kurt Vanhoutte

Astronomy crosses the history of the magic lantern at a decisive point in time. Christiaan Huygens is known as an astronomer for his design of lenses and his discovery of the first of Saturn's moons, but he can also be credited with the invention of the lantern. Huygens' lantern in 1659 contained all the functional elements of the twentieth century slide projector: a box fitted with a set of lenses and a light source to project an enlarged image onto a screen. Huygens could not foresee that a hundred years later his apparatus would become a visual mass medium for the popularization of astronomical knowledge. During the second half of the nineteenth century, lantern shows easily welcomed as many as a thousand participants per evening.

Commercial production is marked by the invention of the 'copper-plate slides' by Philip Carpenter of London. This process entailed printing the outline of astronomical diagrams on glass slides that were subsequently hand colored. By the mid-nineteenth century Carpenter & Westley were accordingly mass-producing the *Compendium of Astronomy*, mahogany framed slides in the familiar 4x7 format.[16] The series and accompanying readings drew large crowds on a daily basis to the Royal Polytechnic Institution, the educational and cultural hub for science and innovation in London. The images follow a mechanical reasoning reminiscent of the orrery, a table-top size model that shows the relative positions and motions of the planets and the moons.[17] Lantern slides moreover added more complex matters such as the causes of twilight, stellar parallax, and the principles of perspective. Itinerant show people travelled the world making use of slides that were all similar as they reflected the accepted state of knowledge. As such, astronomy performance became an important part of the nascent infotainment industry that took place not only in learned environments and lecture halls but also in theatre and opera houses, and that spilled out into public space, the boulevards, and the fairgrounds.

Lantern shows were live, embodied and event-based performances mixing projection with music, rhetoric and interaction. They are the forerunners of our dome-shaped planetariums.[18] 'M. Robin has made a real tableau of astronomy', an old hand reported on one of those shows, 'Nothing is immobile, therein lies the spectacle's charm, and also its merit. The celestial bodies silently follow their paths; meanwhile, their satellites evolve around them. (…) It is a lesson learned while laughing; it's education without effort.' Henri Robin, a magic conjurer turned science demonstrator in Brussels and Paris, indeed reaped great success with his use of mechanical rackwork slides typical for the display of astronomy. A small handle set the constituent parts in motion during (backdrop) projection on a big screen. It is striking that astronomical series invariably also contained beautiful representations of the zodiac and its mythological creatures, thus endorsing the magical appeal of the lantern. The aura did not wane when photographic slides were integrated or when slide production declined in the 1920s. Today, the luminous objects lie dormant in the collections of observatories, schools and museums around the world.

10   Advertising Astronomical Lectures in the Nineteenth Century

Pedro M. P. Raposo

The advertisement in the image from the Adler Planetarium's collections, dating from 1838, promotes a series of six evening lectures about astronomy by

---

[16] Butterworth (2007), p 3.

[17] Vanhoutte (2019), p 145.

[18] Bigg and Vanhoutte (2017).

one Mr. Hall from New York in Lowell. The lectures were to start at 8:00 pm, "affording time for those engaged in the factories to attend." This suggests that the town was Lowell, Massachusetts, which holds a prominent place in the history of the American Industrial Revolution due to its many mills and factories.

The venue was Lowell Street Methodist Church. The lectures would extol the magnificence of God's creation as expressed in the natural world, as evinced by the biblical quote preceding the description of the program ("The Heavens declare the glory of god and the firmament above showeth his handiwork," Psalm 19:1). This was an approach favored by itinerant lecturers in the eighteenth and nineteenth centuries. Another idea that underpins the advertisement is that the planets of the solar system all teemed with life, and that even the sun was inhabited. These were subjects of serious debate among the astronomers and scholars of the time that frequently infused popular publications and events.

The advertisement indicates that some mechanism of projection would be at play: "Lights will be extinguished" and "upwards of 150 diagrams will be presented amid the surrounding darkness." The image illustrating the advertisement, showing what look like phases of an eclipse, a planetary transit, and surface features of celestial bodies as seen through a telescope, were possibly meant to provide an idea of the material that would be projected onto the "large scene, 15 feet square, suspended between the two galleries, affording a view to the whole audience at the same time."

Each of the six lectures, running from Monday to Saturday, focused on specific astronomical objects and phenomena, including: the sun; the moon; comets, asteroids, and eclipses; the primary planets (divided into two lectures); and the fixed stars. Admission at the door cost 12.5 cents, corresponding to an amount in the region of 4–5 US dollars in 2024. There were special prices for the full series and for families, and the public was advised that ""crowded houses are expected, seats should be secured in season."

Although "Astronomy!" was emphasized as the focus of the series, the advertisement also promised to take the audience into the microscopic world by displaying "the foot of a fly as large as an ox, also a portion of the eye of this insect, equally magnified" and "the eye of a point of a Cambric needle, as large and rough as a rock." Expanding on the sensationalist tone of these descriptions, the advert further boasts a view of the "most terrible of all serpents, the Boa Constrictor, in the act of crushing a Wild Beast to death."

It is tantalizing to imagine who attended the lectures, to what extent what was promised in the advertisement was delivered, and what members of the audience would have made of it. The advertisement leaves us with more questions than answers, but the single fact that this piece of ephemera from the 1830s survives is quite remarkable in itself. And it certainly hints at the efforts of nineteenth-century science popularizers to make a stir (and some revenue) with a kind of spectacle whose resonance with what came to be the modern planetarium is paramount (Fig. 6).

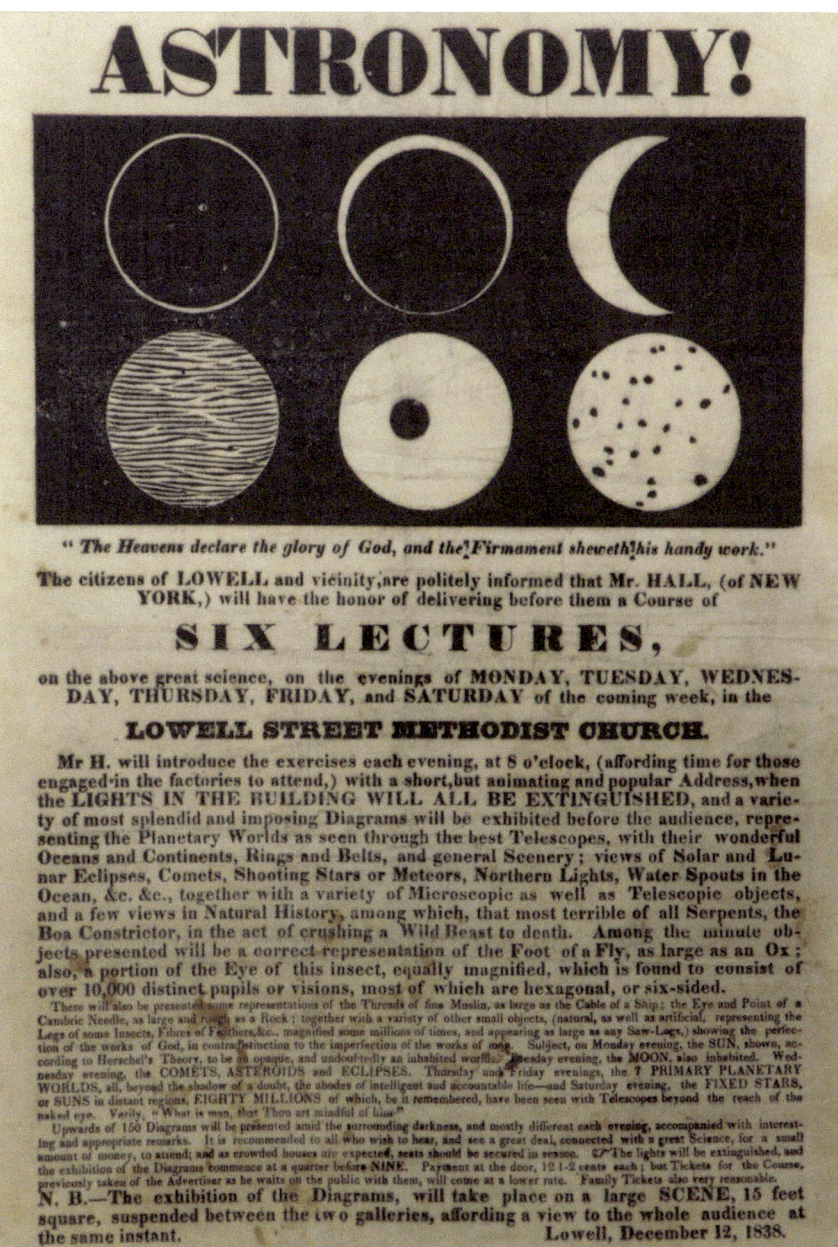

Fig. 6   Advertisement for astronomy lecture series, 1838 (Adler Planetarium collections, P-29)

11   The Copernican Planetarium

Katie Boyce-Jacino

When the Bavarian industrialist Oscar von Miller first conceived of a museum that celebrated German science and technology, he imagined it as an immersive place that would tell the story of German innovation in both the content of the exhibits and the structures of the exhibits themselves. Between von Miller's first plans for the museum in 1903 to its opening in May of 1925, he developed plans for a series of large-scale immersive experiences unlike anything contemporary museums had to offer. In this way, he argued, the museum would serve as a "temple of glory" to German engineering—celebrating past accomplishments and also serving as a place to showcase modern innovations.

The crown jewels of this plan were, without a doubt, the two immersive spaces that dominated the astronomy exhibit. The astronomy wing of the museum occupied three stories directly in the center of the museum building, and was structured as a teleological journey from historical cosmological ignorance to modern astronomical knowledge. Visitors would begin at the bottom level to learn about the dark days of astrology, then ascend to the misguided science of the Ptolemaic system, before climbing the stairs to the enlightened Copernican world. The exhibit ended on a terrace on the topmost floor, where telescopes were set up to allow visitors to actually view the skies above. The two cornerstones of this exhibit were the immersive spaces constructed by von Miller in collaboration with the Carl Zeiss company of Jena that allowed visitors to inhabit both the Ptolemaic and Copernican systems. In naming these two areas, von Miller and his colleagues pulled from the historical tradition of solar system models, and called them Planetariums.

Von Miller developed the idea for the Copernican planetarium first, early in 1913. He envisioned it as a room-sized mechanical orrery, driven by motors, with the planets and sun hanging from the ceiling at relatively to-scale positions from the sun (with the exception of Saturn and Jupiter, whose orbits were shrunk to accommodate the space), and rotated in slight elliptical orbits. (in this, he was inspired by the Eise Eisinga house orrery). The sun hung in the center of the orbits and contained an electric lamp which illuminated the faces of the glass-blown planets. The room, following Sickenberger's suggestion, was a cylindrical chamber twelve meters in diameter and 2.8 meters high; 180 glowing stars studded the wall (an effect created by electric lamps of relative brightness behind small holes in the wall), and the twelve zodiac constellations were painted in gold.[1]

The most distinctive feature of the room was the viewing platform mounted on a rail in the floor that perfectly matched the Earth rail in the ceiling. One visitor at a time could stand on the platform and look through a wide-angle periscope lens. As the track rotated (at a speed of one full rotation every twelve minutes), the sun would remain in the center of the visitor's vision, while the planets would appear to move in and out of view, and occasionally move backwards in retrograde. The planetarium thus demonstrated how the apparent epicyclic motion of the planets was actually caused by the relative position of the earth and the sun.

The Copernican planetarium was an extraordinarily ambitious design, and it took von Miller more than five years to convince the Zeiss company to manufacture it. It was tremendously successful in its time, though its popularity was quickly eclipsed by its more sensational and immersive sibling, the Ptolemaic projection planetarium. When the Deutsches Museum was bombed in the later years of World War II, the Copernican planetarium was destroyed, and was never rebuilt. Zeiss never manufactured another model. Nonetheless, the Copernican planetarium remains an essential piece of the planetarium's history, an early attempt at immersive apparatuses as a way to teach visitors through sensational experiences.

12   The Zeiss Mark I

Katie Boyce-Jacino

What we now know as the Zeiss Mark I, the first projection planetarium, was originally constructed as one of a pair of immersive astronomical apparatuses. Its first name was the Ptolemaic planetarium, and was originally constructed for the Deutsches Museum in Munich as a way to explain the geocentric Ptolemaic system. It was intended as a contrast to the Copernican planetarium, in which viewers would occupy the position of Earth as it orbited around the Sun. But while the Copernican planetarium no longer exists, the Ptolemaic planetarium became an international sensation.

Von Miller's first prototype, developed in collaboration with colleague Max Wolf, proposed a large hollow, 7 m-diameter sphere lined with electric stars, which would rotate around a raised platform representing a fixed Earth position, in which a viewer could stand and look up at the stars orbiting around them. Wolf also added a simplified geared structure within the dome to demonstrate planetary and solar motion. In a curious twist, a nearly identical design was under construction at the same time in Chicago. Neither Atwood nor Wolf had any idea of the others' design, but both were clearly influenced by a much older tradition of rotating globes, stretching back to the Gottorp Globe of the 1650s.

The complexity of the electrical system proposed meant that von Miller could not find a manufacturer willing to undertake the task—even the Zeiss company balked at first. However, after several years of furious back-and-forths, Zeiss engineer Walter Bauersfeld invited von Miller and Wolf to the Zeiss offices in Jena to discuss the idea in person. The main topic of discussion was the logistical challenges of the original design. It was Bauersfeld who first suggested that instead of trying to build a complicated geared system and electric dome, they might construct a modified projection apparatus that could shine the sun and planets along the ecliptic without the cumbersome structure of arms. Almost immediately, Bauersfeld expanded on the idea. As he later recalled, he blurted out, "and why not also the fixed stars?"[8]

The proposal to project everything from a central optical device rather than mount every individual star on the surface of the dome allowed for a dramatic redesign of the entire structure. With all the optics constrained to a projector, which Bauersfeld envisioned as a small sphere mounted on a reflective concave dish on a two-meter high stand in the middle of the room, the room no longer had to rotate.Instead, Bauersfeld suggested a stationary hemispherical dome about nine meters in diameter.

The onset of World War One delayed further work on the apparatus, but by 1919, Bauersfeld began to revisit the design in earnest. By 1921, he had begun to construct a physical prototype of the now-iconic sphere-and-cylinder projector. The sphere was studded with 31 miniature projectors, all of which together could display 4500 stars. Eleven additional projectors at the base of the sphere produced an image of the Milky Way, and an additional thirty projected labels and outlines of constellations (both the Milky Way and constellation projectors could be switched off). In the cylindrical extension, nine separate projectors, each with their own gear shaft, projected celestial bodies—two for the sun, two for the moon (with the capacity to show different lunar phases), and five additional ones belonging to planets (Mercury, Venus, Mars, Jupiter, and Saturn).

Bauersfeld authorized the first test run of the projector in July of 1923, though it took several months of calibration to produce the correct clarity and focus. In September he wrote excitedly to von Miller to let him know that the projector was fully operational. Von Miller rushed out the following day to see for himself.

Finally, in March of 1925, after several more years devoted to refining the machine and perfecting dome construction in the museum building, the Mark I planetarium was shipped carefully to Munich and installed a floor below the Copernican planetarium. It quickly became one of the most popular attractions of the museum, and the focal point for much of the contemporary press on the museum's achievements. According to visitor records, from May 1925 to March 1928, eighty percent of the 2.2 million visitors to the museum came through the planetarium. Zeiss immediately recognized the potential of such a device, but also the technological limitations of it. Bauersfeld and his team thus set about redesigning the projector, resulting in the beloved, instantly recognizable Mark II.

13   The ZEISS Mark II Planetarium Projector

Mike Smail

On May 12, 1930, the Adler Planetarium opened its doors to the public. At the very center of the building, both architecturally and philosophically, was the ZEISS model II planetarium projector. This mechanical marvel projected the night sky, accurately depicting the motions of stars, planets and moons for the education and entertainment of diverse public audiences.

The projector was the invention of a talented team of opticians and engineers at ZEISS in Jena, Germany. Their first planetarium projector, the model I, debuted in 1923, and the first model II was installed in 1926. The Adler's ZEISS model II is noteworthy as it was the first planetarium projector installed outside of Europe. In 1928, Max Adler traveled to Munich, Germany to witness a ZEISS planetarium projector first-hand. He was so impressed by it that he committed to funding a planetarium in Chicago. After 1930, philanthropists and museum directors from across the United States traveled to Chicago to see this incredible device, and by the end of the 1930s, ZEISS model II planetarium projectors could also be found in Philadelphia, Los Angeles, New York, and Pittsburgh.

Stars were projected out of the two large spheres on either end of the instrument. In the middle of each sphere was a 1000 watt lamp. Each lamp's light shone through a series of lenses and 16 star plates, copper discs with small holes carefully hand-punched in patterns that matched the positions of stars in the sky. Light passing through those plates then shone onto a 70′ diameter linen dome, impressively recreating a dark night sky, filled with thousands of stars. The Moon, Sun, and planets originated from a series of projectors in the cages adjacent to the star spheres. Those projectors were geared together to allow the projected objects to move across the planetarium sky in the same complicated patterns traced out by the actual celestial bodies. The projector's center axle contained motors for rotating the stars, moving forward or backward in time, or changing location, to enable audiences to view the sky from a different part of the Earth.

This projector stayed in service for 40 years before being replaced by a ZEISS model VI planetarium projector. From 1970 to 2020, the projector traveled to Mississippi, New York, Louisiana, and Ohio as it cycled between several owners. But it is once again back in Chicago where work is underway to restore it for display at its first home, the Adler Planetarium (Figs. 7 and 8).

Fig. 7 ZEISS model II planetarium projector (Adler Planetarium)

Fig. 8 Adler's first Director, Philip Fox, operates the ZEISS model II for an audience, 1930–1937 (© Kaufmann & Fabry, Adler Planetarium)

14   Evans and Sutherland Digistar 2

Matthew McMahon

Evans and Sutherland, today one of the major developers of planetarium dome technology, was founded in 1968 by two professors in the Computer Science Department of the University of Utah. Professor David Evans and Professor Ivan Sutherland were some of the first to see the potential for computer graphics and founded the company to produce the hardware they needed to further their research. Starting with flight simulators and ship simulators, they also developed display systems for some of the earliest data visualization laboratories in the world.

In 1983 the first of the Digistar planetarium projection systems was installed at the Science Museum of Virginia. It had taken six years to develop the projection system and the technology marked a departure from the opto-mechanical projection planetariums that had been built for sixty years previously. The heart of the Digistar 1 and later Digistar 2 systems was a cathode-ray tube, topped with a phosphor plate onto which the tubes electron beam painted the stars. A single lens which could project a 160-degree image, covering the interior surface of the planetarium dome, finished off the system. The computer at the heart of the system was a VAX-11/780, but this would vary on later models such as the Digistar 2.

Armagh Planetarium was the seventh Digistar 2 system to be installed in the world and replaced the Minolta-Viewlex opto-mechanical projector in the summer of 1995. Digital projectors had many advantages over their older predecessors. They could project vector graphics that were capable of being used to create new and spectacular images on the dome. However, the individual stars did not look as impressive as those made by an opto-mechanical projector, and there were initial issues with the focus on the projector system at the periphery of the dome. These issues have meant that the digital planetarium projector systems never became universal, and many planetaria kept their older Zeiss, GOTO and Spitz systems well into the early 2000's.

Today many planetaria use the best of both worlds, blending opto-mechanical stars with digitally projected special effects and many planetarium projector manufacturers have developed tandem systems that work together to transport the audience in immersive shows. Digistar remains one of the most popular digital planetarium systems and continues to be supported and developed with new features and increasing capability (Fig. 9).

Fig. 9 The Digistar 2, installed in 1995 (Armagh Observatory and Planetarium)

15   GOTO 'Mars' Projector

Matthew McMahon

GOTO was established in 1926 in Tokyo, Japan, by Seize Goto. Originally established as an entry level telescope manufacturer, the business continued to grow steadily. In 1954 the Japanese government passed the 'Enactment of the Science Education Promotion Act' which saw massive investment in scientific equipment in schools across Japan such as microscopes, telescopes and small planetarium projectors. Taking their expertise in optical manufacturing, the company began to provide planetarium projectors for the school market, and by the end of the decade they had acquired over ninety percent of the market share in Japan.

The end of 1950's also saw the release of their flagship planetarium projector, the Model M, or the 'Mars' as it was known in the export market. The Model M-1 was released in 1959 and was designed for use in ten-meter domes. It projected 4500 stars. The design differed from the Zeiss Mark Two's by moving the star balls to the center of the machine, similar to the instrument used at the Morrison Planetarium in the United States. This was to prevent stress on the main body of the instrument, while still giving space for the mounting of planet projectors and special effect projection systems. The GOTO company intended this projector system to catapult them into the global planetaria market, and their second Model M was installed overseas, in the Museum of Art, Science and Industry in Connecticut, in the United States.

The GOTO company had offices in New York, London and Tokyo, and in the summer of 1965, the Armagh Planetarium purchased a 'Mars' or Model M-1 system for their new planetarium. Though Spitz Laboratories had also come very close to winning the contract, when the planetarium opened in 1968 it had a GOTO at the heart of the dome. The GOTO Mars would continue to serve as the primary instrument until 1975, when it was replaced by a Minolta-Viewlex instrument. The Armagh GOTO would eventually be sold to a group in Dublin, in the Republic of Ireland, in 1985. They hoped to create a planetarium for the city, but sadly no dome could be secured and the projector sat unused for a decade. In 1995 it was tragically destroyed in a fire that ripped through the building it was stored in, leaving no trace of this inspiring machine (Fig. 10).

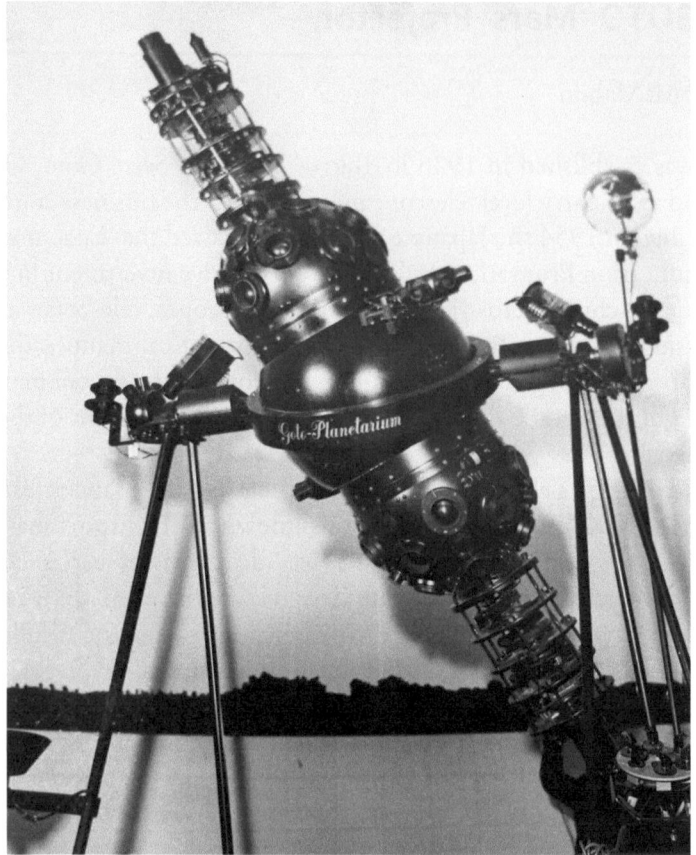

Fig. 10 The Armagh Planetarium GOTO, with the skyline of Armagh surrounding the dome. (Armagh Observatory and Planetarium)

16   The Emmons-HPA Projector

Gary Likert

The fateful day I read the name Richard (Dick) Emmons, the Home Planetarium Association (HPA) was born. For this was the day I received my Starry Messenger in the mail, the old paper astronomy classified ads, and saw one of Emmons' star balls for sale there. I had long dreamed of getting into home planetariums after playing with them as a kid, and this was the watershed event I didn't know I'd been waiting for. I watched with almost involuntary detachment as the check wrote itself and a drilled metal globe was ordered. A week later an unpainted 11-inch metal earth globe appeared on my doorstep. I then took up correspondence with Dick Emmons, and later would meet him at an IPS GLPA conference in Cleveland, Ohio (1997). Along with advice for mounting, housing and using his starball, at that regional meeting he also made one last astonishing suggestion. Perhaps the few of us interested in planetariums at home should begin talking, maybe even via a newsletter. The Home Planetarium Association newsletter and group was born that day, the world's first group in this tiny virtually unknown niche of planetariums at home.

But how was it that Richard Emmons was selling drilled metal star balls to begin with? He and his brother Tom, had marketed a line of 'TSA' projectors to small colleges and other Astro clubs back in the 50 and early 60's. This homemade planetarium featured a carefully hand drilled metal globe, which provided pinhole projection for over 450 accurately positioned stars, plus the Andromeda Galaxy and the Orion Nebula. Its realistic projected sky had permitted a user to recognize all of the bright constellations, and for me it launched me on a journey that lasted for decades, a journey of dome building and starshow giving to the public. In this way, I learned I was following in Dick Emmons footsteps, as he had done the same thing using his prototype projector in his garage under a 14-foot dome. Thousands had been through his planetarium, and my HPA Newsletter soon located several dozen more enthusiasts around the world. A total of 21 quarterly issues would result.

Over the years, with help, I have since uncovered the names of many more lone home planetarium builders going back to the early days of the twentieth century, many through old newspaper searches. It seemed that home planetariums as a hobby was a niche with no name, an invisible pastime, because its participants before my group had been separated by time and space. I then knew what I had to do, and I set about to preserve the Emmons-HPA projector for future generations, but how?

Fortunately, by this time I had heard of Owen Phairis' wonderful Planetarium Museum in Big Bear Lake, CA and as I had moved on to a more detailed star cylinder built by Steven Smith of Douglas, AZ, I made a decision. At this time in the late 90 s, I resolved to donate the Emmons projector to Owen's Museum. It took me a while to get it all packed up and sent to California, but I managed at last, and the Emmons-HPA projector went on to become the only home planetarium in the Museum of Planetariums. The Emmons-HPA projector, once received, was enhanced by Owen and raised, so that it could stand proudly side by side with the Monsters of the Midway, the way larger Zeiss, Minolta, and Spitz machines Owen had painstakingly salvaged and reconstructed for his Museum. He was now giving shows in his facility, as well as renting some of his collection out for major motion pictures! The Emmons-HPA was in good company and would be preserved at last. As I began to self-publish books on my home planetarium experiences, I sent copies to the museum archives as well for preservation along with Owen's vast impressive collection. I wanted to honor Richard Emmons, so I have unofficially dubbed him 'the Father of Home Planetariums'. I hope he would have been pleased.

So this is how a chance acquisition of one of Dick Emmons' last starballs led to meeting with him, which then led to the world's first (and only) home planetarium organization (HPA). The rest as they say is history, and Owen Phairis' Museum of Planetarium is exactly the right place to keep this important legacy alive! It is indeed an oddity among the oddities of the museum.

I thank Owen and everyone else who over the years has played a part in this legacy (Fig. 11).

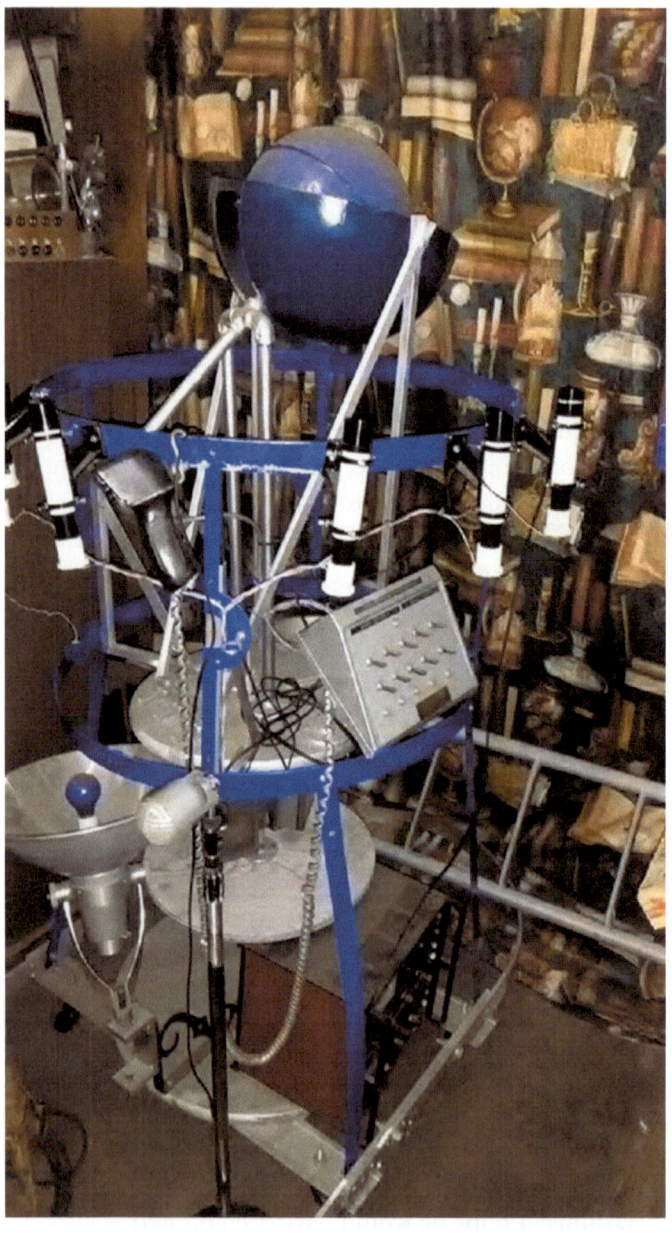

Fig. 11 The Emmon's Projector, currently preserved in the Planetarium Museum (Gary Likert)

17   Light Pointer as Showstopper

Helen Ahner

The planetarium brings the stars within reach—this promise has been attracting people under the dome since the first public planetarium shows in the summer of 1924. To fulfill this promise, it took far more than the planetarium projector showcasing the night sky. One technical gadget in particular mediated between the celestial projection and the earthly audience: the light pointer.

The device projected a small light arrow onto the planetarium dome, which lecturers could use to point precisely to the spot in the projection they were talking about. The light pointer overcame the distance of often 5 to 10 meters between the speaker's desk and the dome and solved a didactic problem encountered by astronomy popularizers: With the light pointer's help, there was no longer any confusion as to exactly which celestial body was being discussed—a problem that was not so easy to solve when observing the starry sky outside the planetarium.

It cannot be determined conclusively whether the light pointer was developed by the Zeiss company *for* the planetarium or whether its introduction merely *coincided with* the invention of the planetarium. What is certain is that the first planetarium projector that the Zeiss company designed for the Deutsches Museum in Munich was already equipped with such a gadget. Press reports on the early planetarium shows also indicate that many guests first became acquainted with such a device under the planetarium dome and considered it a sensation worth mentioning.

The planetarium visitors were enthusiastic about the technical gesture enabled by the light pointer. For them, the device was more than just an optical cane. Some of them recognized in the light pointer a herald of future technical ventures into space. If it was possible to intervene in celestial events and touch the stars in the planetarium using technology and science, how long could it be before people could also overcome the distance to the moon outside the dome.

One of the reasons why the light pointer had such a stimulating effect on the imagination of planetarium spectators was because it staged and visualized a connection between above and below, lecture and projection, body and technology, humankind and stars. In newspaper articles about the planetarium from the 1920s, pointing with the device was imagined as playing tag with the constellations, chasing the celestial bodies or teasingly touching the stars. The lively physicality of these gestures in particular refers to the sensual and corporeal appeal of the device. Product photographs and sketches also highlight the fact that the light pointer had a special

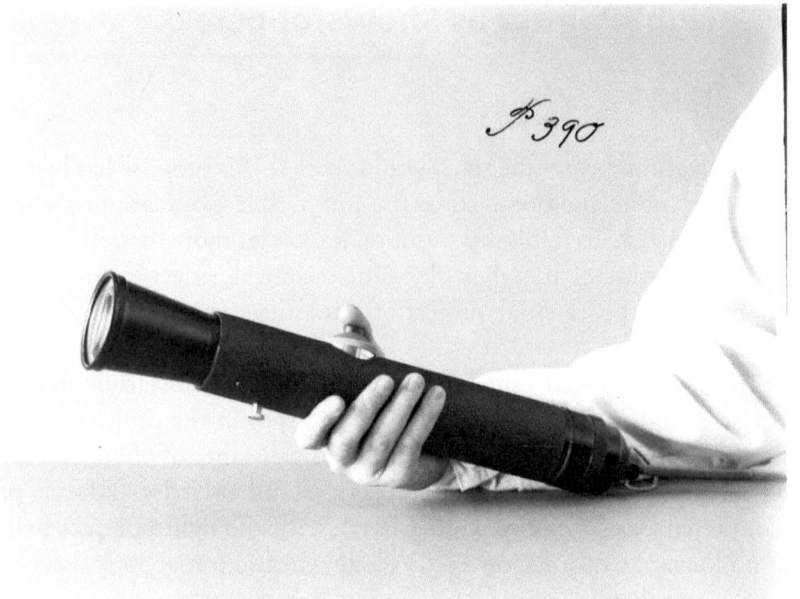

Fig. 12 Early light pointer from Zeiss (BL 06198, Zeiss Archives)

connection to the human body: in most of the illustrations, the apparatus is shown in conjunction with a human hand and thus depicted as a technical extension of physical reach.

As a didactic aid, the light pointer is still one of the most prominent technical agents in planetarium shows and one of the presenters' favorite accessories. In addition to the educational pointing gesture that it enables, it is not least the playful interaction with the distant celestial bodies and the touching of the seemingly untouchable that ennoble it as such an exciting artifact (Fig. 12).

18   Public Telescope at Armagh Planetarium

Matthew McMahon

The public do not usually know the difference between an observatory, and a planetarium. They are usually quite surprised to discover they can not only be two very different buildings, but also can be surprised to discover one usually exists without the other, though there are nearly as many exceptions to this rule. Both institutions have a dome, or several, and something to do with the stars and galaxies above our heads. However it is not a rule that a planetarium must also have a telescope of their own.

The Armagh Planetarium was first discussed by the seventh Director of the Armagh Observatory, Dr Eric M. Lindsay, in 1943. From the start he envisioned that the public, having seen the illusionary sky on the dome, would want to then see "the real thing" through a telescope. However the funding provided by the Northern Ireland government in 1965 would only cover the building and projector, with no funds left for a public telescope. However in the early 1970's, with the initial success of the Armagh Planetarium and a dynamic Director at the head of it, Terence Murtagh, the government agreed to fund both a new exhibition area, and a public telescope.

Since the 1950's the Armagh Observatory had used the Grubb 10-Inch refractor (built in 1885) as both a scientific research instrument, and for public viewing. The mechanical clock drive was still reliable, and the instrument was well balanced and easy to use. To keep costs of the new expansion down it was initially planned that the Grubb would be dismantled and moved to a new dome located closer to the Armagh Planetarium. However tragedy struck when Dr Lindsay passed away suddenly during the summer of 1974. He was replaced by Dr Ernst Öpik as Acting Director, who blocked the move of the Grubb on the grounds that the new dome which had been built was not sufficiently stable for scientific research to continue with the instrument. It was reerected at the Observatory while Terence Murtagh was able to secure additional funding for a modern telescope.

The new telescope, a sixteen inch reflector, was operational for the first winter of observation in 1975. It was operated by Patrick Corvan, a local man who had first become interested in astronomy after a visit to the Observatory in 1953. In 1980 a house was provided onsite for him to live with his family. A night of observation would start with the moon and the public were encouraged to discover for themselves where objects lay in the sky, moving from planets, to particularly interesting clusters and galaxies. Patrick would go on to operate the telescope for the public for thirty years, before he retired in 2005 (Fig. 13).

Fig. 13 Patrick Corvan observing at the telescope during his retirement in 2005. (Armagh Observatory and Planetarium)

19   The Jim Kaler Home Planetarium

David C. Leake

The William M. Staerkel Planetarium at Parkland College in Champaign, Illinois was fortunate to be the recipient of a homebuilt planetarium constructed by Dr. James B. Kaler when he was a curious youth in Albany, New York.

Kaler called our attention to an article he published in the Proceedings of the Astronomical League General Convention, held at the Carnegie Institution in Washington, D.C. on September 4–7, 1953. Jim was 14 years old at the time.

In the article (a copy of which is displayed next to his planetarium in the Staerkel lobby), Jim details how he used a hammer and a nail to punch holes of different diameters in a Crisco can. Jim drew celestial coordinates on the can, with the top being the north celestial pole. There are approximately 500 stars projected down to 4th magnitude. The Milky Way projector was another Crisco can with a cut lengthwise and then clear plastic was mounted over the strip, and painted for texture. A third Crisco can became a meteor projector. A piece of paper with a hole in it passed over a slit in the can yielding a moving dot in the "sky." Flashlight bulbs provide illumination for each can. The entire projector was created by Jim, his father Earl, and a family friend, Jack Gilliland.

We displayed the home planetarium beneath a white umbrella hanging from the ceiling. When turned inside-out it acted as a small dome. The unit was rewired to accept D-batteries but the original rotary switches remain on the device.

Dr. Kaler would go on to earn an A.B. from the University of Michigan and a Ph.D. from UCLA in 1964. He was a staff astronomer at the University of Illinois from 1964 until his retirement in 2003, during which time he penned over 450 publications and 19 books.

When the Staerkel Planetarium hosted the Great Lakes Planetarium Association (GLPA) Conference in 1989, then director Dave Linton needed a speaker. He jokingly said Jim was "local and he was cheap." He gave the first Astronomy Update talk to the conference, and feedback was so positive that Jim was asked back to provide the update talk every year through 2008.

Dr. Kaler was an annual participant in the Staerkel Planetarium's "World of Science" lecture series and the series now carries his name. GLPA awarded Jim both the "Fellow" award and an Honorary Life Membership, and he was featured in the GLPA fulldome planetarium show, "The Stargazer."

Dr. James B. Kaler passed away on November 26, 2022 at the age of 83. Building a planetarium as a teenager was simply the first step in Jim's long history as a great friend to the planetarium profession (Fig. 14).

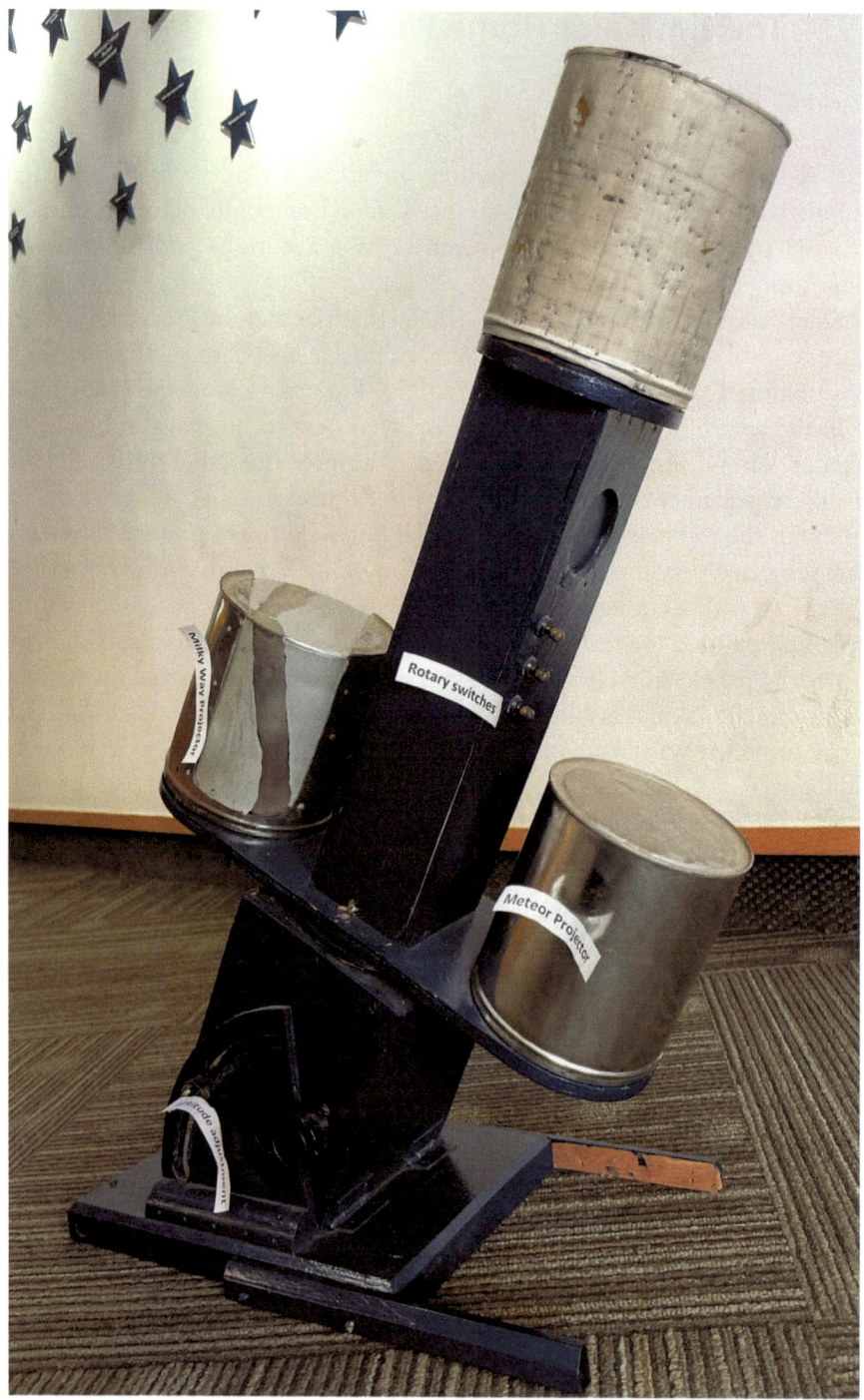

Fig. 14 Dr. James B. Kaler's handmade planetarium (William M. Staerkel Planetarium, Parkland College)

20   LSS Planetariums

Lionel Ruiz

Before the 2000s, making a planetarium projector required mechanical skills, a certain skill but above all a lot of time for its design and production. Otherwise, you had to have a well-filled wallet, allowing you to afford these star projection systems, whether with or without planets. The shift to digital projection systems for planetariums that took place in the 2000s facilitated access to full-dome digital projection for small budgets. Projection systems with the help of a quarter spherical mirror were beginning to emerge but were only properly usable for the projection of films because the rendering of a starry sky did not really make it possible to snatch wows of satisfaction from the spectators. The small planetariums in France shunned the fact of projecting only films and were more interested in having an exploitable starry sky, a key point in a session of discovery of the night sky. In 2008, Yves Lhoumeau developed, using camera lenses, a fisheye optical complement allowing, using a conventional video projector, to project a luminous image over the entire vault of a small planetarium dome. The concept was improved with the help and advice of Lionel Ruiz in order to bend the optics so as to prevent dust from attaching to certain critical parts and to optimize the airflow in a position intended for it. With the help of a fork of the famous Stellarium open-source software, specifically adapted to the needs of planetariums, it is now possible to assemble a complete system for less than US$3000 (computer + software + video projector + fisheye optics). Fifteen years later, the number of digital planetariums in France has continued to grow and the dynamic instilled by this system with their users during regular meetings brings that it is deployed in almost all of the small installations present in the French territory. This system has been emulated around the world with more than 300 planetariums to date declared as LSS, and allows many people to enjoy sessions of all kinds. Now, the only limit we have left is that of our imagination.

All the details for its construction are available online by typing the keywords "LSS" and "Planetarium" on the Web (Figs. 15 and 16).

Fig. 15 Image of the inside of a dome in an LSS Planetarium (Lionel Ruiz)

Fig. 16 Fish-eye lens projection system (Lionel Ruiz)

21   Armand Spitz's "Soap Can"

Michael McConville

For most of the twentieth century, the ubiquitous tool for planetariums in the United States was a Spitz A-series star projector. What a Spitz projector might have lacked in refinement or capability compared to the ZEISS opto-mechanical star projectors from Germany, it more than made up for in value—Spitz projectors, efficiently made, were many times less expensive than their German-made counterparts.

But before the dodecahedron of the A-1 helped to make Armand Spitz the progenitor of the "American Planetarium" in 1947, a lowly laundry soap can would serve as the first step for realizing the Philadelphia-based meteorologist and science communicator's dream of a "projector for everyone." An unassuming cylinder produced by the Crystal Soap and Chemical Company, Inc., of Philadelphia, the prototype pinhole projector stands about 10 inches tall, and roughly seven-and-a-half inches in diameter. The can showcases a distinctive patina after nine decades of handling, with a faded adhesive label that reflects the years of use by Armand Spitz—for the first years of his eponymous company, Spitz would take the Soap Can with him to demonstrations and exhibitions as a testament to the evolution of the company's projectors.

Even a cursory glance at the exterior of the Soap Can reveals many of the most prominent northern hemisphere star patterns. The "top" of the projector—the original bottom base of the can—is home to the northern circumpolar constellations, with Polaris punched dead-center in the starfield. Ursa Minor and Cassiopeia are featured in their entirety, with the brighter stars of Draco wrapping around the celestial pole and the stars of the Big Dipper prominently drilled into the top part of the can's side. Spitz's starfield is by no means accurate, particularly in the size (and therefore, apparent brightness) of the stars themselves, but does a fine job of representing the general spacing, size, and orientation of the constellations relative to one another.

The sides of the pinhole projector are replete with the sorts of star patterns and constellations that would be easily recognizable to just about any audience—Orion, the Great Square of Pegasus, the stars of the Summer Triangle, and the constellations of Sagittarius and Scorpius. The southern constellations just skirt the "horizon" at the bottom of the Soap Can, matching their visibility in the sky during the hot and humid Philadelphia summers.

As might be expected with a proof-of-concept implementation like this, the quality of the starfield projection did not compare with contemporary opto-mechanical star machines, like the ZEISS or Korkosz installations throughout the United States that used precision-drilled star plates, focusable lenses for brighter stars, and intricate internal gearings for precision movement of the night sky. But this remarkably simple idea would eventually lead to thousands of Spitz star projectors in domes around the world—democratizing access to the beauty of the night sky for millions more people (Figs. 17 and 18).

Fig. 17   The Big Dipper as seen on the soap can projector (Cosm/Spitz Archives)

Fig. 18 Armand Spitz with the Can Projector (Cosm/Spitz Archives)

22   Spitz B & C Planetarium Projectors

Garry F. Beckstrom

After the successful introduction of the Series A star projectors for small plan-etariums, Spitz Laboratories, as it was known then, designed a projector for large planetarium domes designated the Spitz B. Three Spitz B projectors were constructed before the Spitz STP was later developed for larger domes. Spitz B projectors were installed at the Montevideo Planetarium in Montevideo, Uruguay, 1955; the Robert T. Longway Planetarium, Flint, Michigan, U.S.A. 1958; and the U.S. Air Force Academy Planetarium, Colorado Springs, Colorado, U.S.A. 1959.

As was true with the Series A Spitz projectors, each of the Spitz B projectors was slightly different as improvements were made with each one. The dumbbell-shaped projector was 11.5 feet long and weighed slightly over half a ton. It was suspended from the dome by steel cables and anchored to the floor by similar cables. The concept was that without a huge mount, the audience would be able to see all around the instrument as it appeared to float in the darkened planetarium.

Stars of magnitude 2.0 and brighter were projected by individual lens systems on the 36-inch diameter star hemispheres while the rest were produced by tiny holes. The faintest star shown was of magnitude 5.8 produced by a 0.0135 inch diameter hole. The extreme sharpness of the star images was a result of the concentrated brilliance of zirconium-arc lamps located at the ends of the hemispheres, reflecting off half-sphere mirrors inside. Motorized Suns, phasing Moon and planet projectors occupied the cages between the hemispheres and were belt and pulley driven. Auxiliary projectors for coordinates, meridian, precession circle, constellation outlines and other special effects were also attached to the projector.

The Spitz B could simulate an entire day in as little as one minute and run annual motion through a year in 12 seconds, as well as demonstrating precession. Of special note is that the Spitz B projector at the Montevideo Planetarium was in use until 2018. In addition, the Spitz B hemispheres from the Air Force Academy Planetarium were later modified by Spitz, Inc. and used as a special star projector at a New York City nightclub.

A similar projector—half the size of the Spitz B—but of the same design and capabilities, was manufactured for the Minneapolis Public Library Planetarium in Minneapolis, Minnesota U.S.A. and was designated the Spitz C. It was the only Spitz C built. Maxine Haarstick, then director of the Minneapolis Planetarium was having a 40-foot planetarium constructed, too small for the Spitz B. Legend has it that she convinced Armand Spitz to build her this special projector, as she liked what she saw of the Spitz B. The Spitz C operated in Minneapolis until 2002 when that building was closed (Figs. 19 and 20).

Fig. 19 Spitz B Planetarium Projector, Robert T. Longway Planetarium, Flint, Michigan U.S.A. In operation 1958–2001. (Garry F. Beckstrom)

Fig. 20 Spitz C Planetarium Projector, Minneapolis Public Library Planetarium, Minneapolis, Minnesota U.S.A. In operation 1960–2002. (Garry F. Beckstrom)

23   Spitz Junior Planetarium

Mike Smail

Armand Spitz was renowned for his life's work to develop small, affordable planetarium projectors, that were within the reach of smaller museums, educational facilities, and the general public. So it should come as no surprise that the first widely produced and distributed home planetarium would bear his name.

In the early 1950s, Armand was giving a public demonstration of his first planetarium projector, the Spitz Model A. In the audience was Thomas K. Liversidge, the CEO of Harmonic Reed, a Philadelphia company that produced toy musical instruments. After seeing the demonstration, Liversidge wondered if the projector could be simplified into a toy version. Armand was intrigued by the idea, and the two companies teamed up to develop the Spitz Junior Planetarium.

Harmonic Reed's engineering team set to work designing and fabricating the starball. Instead of using Spitz's dodecahedron shape, Harmonic Reed solved the incredible challenge of producing a spherical, plastic starball with approximately 300 holes punched into its surface. The finished design consisted of two hemispheres, joined by a rubber gasket. A lamp mounted inside the starball projected the stars out through the surface holes. The projector also featured altitude control so the viewer could view the stars from anywhere in the Northern Hemisphere, hand control of diurnal motion, and a rheostat to control the brightness of the lamp. Completing the package was a cardboard apparatus for adapting 'most any "d" cell flashlight so that it can be used as a pointer' and an illustrated 32-page booklet written by Armand Spitz. This book both explained setup and operation of the projector, and contained star maps and other information to identify celestial objects.

The Spitz Junior Planetarium was released in 1954 and cost $14.95 (about $175 in 2024 dollars). The next year, the Sky Zoo was unveiled. With the exception of the starball, this projector was almost a carbon copy of the Spitz Junior. The Sky Zoo starball was covered in translucent constellation artwork and was designed to be used in tandem with the Spitz Junior: one projector for the stars, and another to project artwork. The Sky Zoo projector also included a wired, arrow pointer. For budget-conscious folks who already had a Spitz Junior, the Sky Zoo starball was also available by itself for only $3.98 (about $46 in 2024). Between 1955 and 1958, a few thousand Sky Zoos were produced, before being discontinued.

One final version of the projector was released in 1958, the Portable Jr. Planetarium. This projector featured the original starball design mounted to a slightly smaller base, ran on D batteries and came with a battery-powered arrow pointer. It cost $10.00 (about $115 in 2024 dollars), and touted the ability to 'project over 60 constellations on any ceiling'. Interestingly enough, the Portable Jr. Planetarium included no mention of the name 'Spitz'; it was only branded with the 'Harmonic Reed' name.

1958 also brought significant improvements to the original Spitz Junior Planetarium. It now came with the wired arrow pointer first included with the Sky Zoo, and supplementary constellation, Saturn, satellite and eclipse projectors. Harmonic Reed also sold some of these ancillary projectors as a stand-alone product under the name 'Starmaster Astronomy Set'. Interested product owners could send in a form, and pay a small fee, to join the Junior Planetarium Association, for which they received a membership card and pin. Finally, a new Southern Hemisphere Projector was released, model 3500 (the Northern Hemisphere version was model 3000).

Despite the improvements and perks of this toy planetarium, by the early 1970s, the lustre of the Space Race on the American populace had worn off, and the entire line was discontinued. In just under 20 years, about a million of these projectors were sold. Those incredible sales numbers means that, even today, the Spitz Junior Planetarium is still accessible through eBay and other online markets, giving an entirely new generation the chance to be enthralled by an object described as 'the scientific toy of the century' (Figs. 21 and 22).

Fig. 21 A membership pin from the Junior Planetarium Association (Mike Smail)

Fig. 22   An early Spitz Junior Planetarium, with box (Mike Smail)

24   5000 Glowing Stars in the North Woods

Mike Smail

The Northwoods of Wisconsin are a unique place, home to the World's Largest Black Bear, the World's Largest Soup Kettle, the World's Largest Penny and the World's Largest Mechanical Globe Planetarium. The last of those items is due entirely to the tenacity and unending grit of Frank Kovac Jr.

Frank grew up in Chicago, and like many people, was fascinated by his visits to the Adler Planetarium. In Frank's own words 'the highlight of my visits to the Adler were the wonderful sky shows, which eventually led me to where I am today.' Young Frank consumed everything he could find related to astronomy. As a teenager, his father gave him his first telescope, a small 60 mm reflector. A few years later, he built his own 10 inch reflector. He stuck luminescent stars on his bedroom ceiling, and painted a sizable mural of Saturn on his bedroom wall. At age 15, Frank's family visited the Northwoods, and he saw the aurora borealis.

After high school, and a stint in the Air Force (the closest Frank thought he could ever get to being an astronaut), Frank found his way back to the Northwoods, and their beautiful dark skies. He bought some land outside of Monico, Wisconsin, and built an observatory to share his love of the universe with anybody who visited. But even though these rural skies were dark and still, they were often cloudy. And it was the frequent cloud cover that finally convinced Frank that he needed to build a planetarium, so that people could see the stars, even when it was cloudy.

Frank's idea was to build a large globe and paint stars on the inside of it. A tilted globe that matched his latitude could turn around the audience, allowing a view of the sky at any time of year. Unknown to Frank at the time, three other mechanical globe planetariums had been built throughout history, including one in his hometown of Chicago, some 80 years prior.

Designs led to experimentation. The earliest version of Frank's planetarium was a geodesic structure of PVC pipes, but it crumbled in the cold Northwoods' winter. For the next version, Frank realized that he needed to construct his planetarium sphere out of sturdier materials inside. This next design would be plywood, glued into 24 large curving sections, with metal panels for support. They would be assembled into a 22 foot diameter globe with a base ring that would rest on rubber wheels affording the diurnal motion of the heavenly globe. But after four years of construction (Frank worked in the evenings and weekends, outside of his day job at a local paper mill), the globe fell as Frank was attempting to hoist it onto the base ring. Despite almost being crushed by the falling globe, Frank soldiered on.

In analyzing the failure of this design, Frank determined that replacing his wooden supports with steel, particularly at the top of the globe, would help prevent another collapse. This work was also being done without blueprints, it was all based on sketches and simple shop math. This led to a significant amount of trial and error along the way, but by the early 2000s, the two-ton sphere was assembled, mounted, and ready for painting.

Research led Frank to an 18-hour luminous paint product for his stars. Each star hand painted by Frank, with constant consultation of Norton's Star Atlas, and frequent trips outside to the night sky to confirm placement and relevant brightness. Brighter stars were painted slightly larger. And if he made a star too bright, Frank would mix in white paint to dim it down. A broader sponge brought the Milky Way and deep sky objects to life. Frank's sky spread across 312 styrene panels affixed to the inside of his planetarium globe. The end result was a little over 5000 glowing stars, accurately depicting a clear dark sky as seen outside the planetarium.

Several thousand people now visit the Kovac Planetarium each year. Frank is the only employee, welcoming and presenting the stars to each one of them. He's rebuilding the observatory on the property, and hopes to continue growing his exhibits and guest capacity. If you're ever in the area, it's well worth the stop. Not just to experience the wonder of the world's largest mechanical planetarium, but to experience the wonder of Frank Kovac Jr (Fig. 23).

Fig. 23   The Kovac Planetarium (Mike Smail)

25   San Francisco's Home-Built Planetarium Projector

Bing Quock

By the time optical projection planetariums came to the United States, the iconic Zeiss Optical Company was still the world's only manufacturer of large star projectors. Between 1930 and 1949, six large Zeiss-equipped domes opened in the U.S. Several instruments for domes under 15 meters (50 feet) were produced domestically, including the Rosecrucian planetarium projector (1936) in San Jose, California, and the Korkosz projector (1937) in Springfield, Massachusetts, but America had not yet produced an instrument for a larger planetarium dome.

Enter G Dallas Hanna, a geologist at the California Academy of Sciences in San Francisco, California, who had led a team of optical workers as part of the national volunteer effort during World War II. Immediately following that conflict, projectors were not available from Zeiss, but Hanna felt that the skilled group he had gathered possessed the expertise to produce a planetarium projector competitive with the German-made instrument. A design was proposed in the form of a sketch by Russell Porter—an innovator in amateur telescope-making involved in the design of the Hale Telescope at Palomar Observatory—helped convince the Academy's Board of Trustees that the project was achievable.

Hanna and machinist Albert Getten decided against the Zeiss "dumbbell" design and redistributed the instrument's mass by moving the heavy starballs closer to the central axis. This improved the balance and allowed for smaller, quieter motors. Another innovation was in the way the star images were produced. Instead of drilling round holes through sheets of metal, Hanna opted to utilize grains of carborundum—an abrasive grit used to polish lenses. These were sorted by size, with larger grains representing brighter stars, and would represent 3800 stars. Although fewer than produced by the Zeiss instrument, this number was thought to look more natural. The thousands of sand-like grains were carefully hand-placed onto the flat surfaces of 32 large condenser lenses—a job that took optical technician Frances Greeby six months to complete. After the grains were positioned, each lens was placed in a vacuum chamber and coated with a layer of vaporized aluminum. When the carborundum was gently brushed away, the tiny openings they left in the aluminum projected stars that were shaped like carborundum grains and were thought to look more realistic than round holes.

After opening in 1952, the Alexander F. Morrison Planetarium was visited by planetarians who praised the quality of its starfield—among them Wagner Schlesinger, Director of Chicago's Zeiss-equipped Adler Planetarium. Several years later, Japanese industrialist Seizo Goto, interested in manufacturing planetarium projectors, evaluated different designs across the U.S., and favored the Morrison model—a configuration that the Goto Corporation emulated for several decades.

After 51 years, the Morrison projector was retired at the end of 2003, when the California Academy of Sciences was temporarily closed for reconstruction. Although no longer in use, the Academy Projector remains fondly remembered by long-time visitors and staff who were able to view or work with this remarkable instrument.

26   The Historical Zeiss Projector of the Planetarium of Rome

Stefano Giovanardi

Rome was the first city to open a Planetarium outside of the German speaking Countries, in 1928. The Planetarium of Rome has a remarkable history, which began when Germany donated to Italy a Zeiss Mark II projector, as a partial compensation for the damages of World War I. The projector was positioned inside a ancient Roman building, the Octagonal Hall, that was part of the thermal baths of Diocletian. Under a Roman vault, built in stone and bricks 2000 years ago, a metallic dome was installed, with a geodetic structure 19 meters in diameter, to serve as the celestial sphere for the astronomical projection. A dome inside the dome, connecting along the same curvature centuries of history, through the timeless beauty of the starry sky.

The word "PLANETARIO" is still carved in capital letters in the travertino stone above the entrance of the Octagonal Hall, as well as the final verse of Dante's Divine Comedy, "l'amor che move il Sole e l'altre stelle" (love that moves the Sun and the other stars), marking the entrance door to the Planetarium hall.

On October 28, 1928 the Planetarium was inaugurated at the presence of Mussolini, and quickly became an attraction for visitors from all over the world, including royal families, such as the kings of Italy and Belgium, cardinals, dukes and other celebrities.

The projection consisted of over 5000 stars from both celestial hemispheres, and adopted ingenuous technical solutions to represent the twilight effect, the Moon and its phases, the movement of the planets, the Milky Way, including mechanical eyelids covering the stars below the horizon, and uncovering them—simply by gravity—as they rose throughout the night, following the projector rotation. Three different rotation axes allowed the projector to simulate the precession effect and the diurnal motion of the sky even as seen from different planets.

The Zeiss Mark II projector gave light to the stars at Rome Planetarium until 1981, when the Planetarium was closed and the projector disassembled and stored in boxes. It was reassembled only 23 years later, when a new Planetarium opened in Rome at the current location, the Museum of Roman Civilization: the Zeiss Mark II is now on display in the Planetarium entrance hall. It is the only integral existing model of its kind (although not operational), as many similar projectors in Germany were damaged or destroyed by World War II.

Being the first model to display both celestial hemispheres, the Mark II introduced the classic two lobe structure, becoming a true icon of Planetarism. Ironically, because of its towering, elegant and unusual design, over the years it has been creatively labeled in the museum floor plan as a "telescope" or a "satellite".

27 The Fleischman Atmospherium-Planetarium Projector

Cora Braun

Up until the 1960s, planetariums were almost exclusively occupied by the star projector itself and its auxiliary projectors for planets and the moon. Additional slide projectors were also used to illustrate the lectures with further images. However, additional projectors always covered only a part of the dome, and— if they were not masked—stood out with a frame that separated them from

Fig. 24 A postcard with a 1960s photo by Doc Kaminski showing Richard Norton giving a presentation with the Atmospherium projector at the Fleischman Atmospherium Planetarium. On the back is noted: "Storms and stars, tornadoes and galaxies—for the first time the audiences can witness both the day and the night sky. A unique projection system reveals the mysteries of weather and the universe within the dome-theater" (Department of Special Collections and University Archives, University of Nevada, Reno)

the surroundings. Thus, for a long time, dome-filling, immersive images could only be produced with a star projector itself and were therefore restricted to astronomy-related topics.

This changed with the opening of the "Atmospherium-Planetarium" at the University of Nevada, Reno, in November 1963: The institute's director at the time, Dr. Wendall A. Mordy, realized that by building a planetarium he could also advance his research on various atmospheric phenomena such as cloud formation. He convinced donors to increase funding so that an "Atmospherium" could be built in addition to the planetarium.

The goal of the Atmospherium projector was that frameless images could be displayed over the full 360° x 180° of the planetarium dome to show various weather phenomena. The viewer should be placed within the center of the activity. The hitherto so typical representation of the night sky at the planetarium was to be extended by the realistic representation of the day sky.

Curator O. Richard Norton used a 35 mm film projector with a special fisheye lens that could project a 160° angle. By recessing the Atmospherium projector into the floor of the planetarium, the full 180° of the dome could be projected. A Spitz A3P projector was also installed in Reno. To keep it out of the way when the Atmospherium was projecting, the A3P was mounted on a rail system so that it could be moved to the north of the dome and lowered into the floor with an elevator if needed.

In addition to the fisheye projector, a corresponding 35-mm fisheye camera was built to take pictures that could be projected inside the dome. Primarily time-lapse recordings were made, which were later played back faster in the dome, so that, for example, the formation of thunderstorms, frontal systems and cyclonic storms could be observed. During the experimental recordings, Norton quickly discovered that the camera system was also suitable for simulated spacecraft approaches and Earth orbits. Furthermore, its use in oceanography also seemed promising: the camera was lowered into shallow waters in a Plexiglas box to capture dome-filled images of various seafloors, particularly their fauna and flora.

However, this projector did not catch on for several reasons and remained a unique specimen for the next decade: In contrast, conventional star projectors provided a fully developed, technical and educational system whereas the Atmospherium was merely an experimental projector. Content production required immense technical, financial and human resources and there were no ready-made "educational packages" to offer planetariums. Similar, less expensive and more advanced Allsky projectors began to spread from 1969 onwards.

The development of Allsky photography paved the way for the planetarium to create immersive images beyond the star projector and open up possibilities to extend the immersive planetarium experience to other topics.

28   Not any Gel to Color the Stars

Carole Homberg

Back in the fall of 1987, I was at my first job, right out of school. I was at my very first job, as an academic staff member at the University of Wisconsin-Oshkosh. One of the most exciting parts of this job was to be the person in charge of the Buckstaff Planetarium, which was housed in a small building just outside the science building. At that time, it contained a Spitz A3PR and either two or three slide projectors. I remember that it did have a wheelchair lift, but it did not have air conditioning, and its metal roof made the building uncomfortably warm during the summer months. One of the physics faculty, John Evans, had been in charge of the planetarium previously, and showed me how to work the controls.

In June 1988, I attended my very first planetarium conference, in Richmond, Virginia, where I saw "The Mars Show" by Loch Ness Productions. Upon my return to Wisconsin, I felt ready to give my first public show. It would be about Mars. It would be a live show, as I either didn't have the capability to make recordings, or I didn't know how to use the equipment. I read Gerald Mallon's column in *the Planetarian* journals and one of his articles dealt with a lesson about Mars. Perfect for me!

One of the interactive portions of Dr. Mallon's Mars planetarium show was to show 3 red star-like objects in the planetarium sky—two stars and the planet Mars—then advance time a few weeks and ask the audience which object moved. But when I practiced my technique, I found that the Spitz planet Mars was much more red than the stars. The color difference was an immediate give away.

I wrote either a letter or an email to Dr. Mallon, explaining my problem. He replied for me to use a gel to make the stars and the planet the same color. While I did buy a tube of red cinnamon-flavored tooth gel, I could not bring myself to smear a little onto the Spitz star ball.

At another planetarium that I worked at a few years later, I found a Wess Gel Kit and that is how I discovered that colored plastic sheets that one places over a light source were called gels. Often when slides were made, one used Kodalith film, especially those that had words rather than pictures. Kodalith produced slide film that had clear images on a black background. To create color on the images, one would sandwich a gel of the desired color into the slide mount.

29   35 mm Astronomical Slides

Terence Murtagh

Slide projection and the planetarium have a long history together, and significant technological overlap. From the 1950's and 60's the glass plate projector was a common part of a planetarium lecturers toolbox. However, the 35 mm slide projector prompted a leap in quality, and an accessible price point that made it an attractive upgrade to many planetariums.

The advent of full color, scientifically accurate photographs of distant galaxies and nebula was an example of how planetaria diversified into many other forms of media. Prior to their development many of the images that were shown to the public of the cosmos were either artistic renditions or black and white copies of glass plates. The artistic renditions were of excellent artistic quality, but there was an acknowledged gap of authenticity that both scientists and the public wanted to cross.

Until the advent of the Armagh slide sets only a handful of colour astronomical images had been available. First announced by a two-page advertisement in *New Scientist* in the early 1980's these 35 mm slides were the result of collaboration between cutting edge astronomers and innovative planetarians.

Whilst in Australia the Director of Armagh Planetarium, Terence Murtagh, made contact with the famous astronomer David Malin. With the help of the magnificent imagery from the Anglo-Australian Telescope, and Schmidt Camera some of the finest 35 mm slides were made. The masters of each set were carefully produced by David Malin and sent to Terence for production.

For many professional astronomers, lecturers and students, these slides were the first time they had ever heard of Armagh Planetarium. The slides were sold through mail order and Terence established international networks with other planetariums all over the world to distribute them.

What made these images so groundbreaking was that David used three different photographic plates, red, green and blue of each image. By recombining these plates, he was able to make images that were truly unique and had scientifically accurate colour. Quickly they became an essential tool in university lecture theatres around the globe. For lecturers these slides met with high praise and were quickly integrated into their own astronomical public outreach as well as the education of the next generation of astronomers.

Not to be confined to the lecture hall these full color images were also projected onto planetarium domes, though still in a rectangular "window" format. However, these slides gave the public a visual understanding of the cosmos that had not been possible before.

Later David provided high quality images which were converted to a 360 degree all sky projection format provided by Sky Skan Inc. Covering the whole dome these images was extraordinarily impactful (Fig. 25).

Fig. 25 Advertisement for the 35 mm slides for sale (Armagh Observatory and Planetarium)

30   Special Effects in the Dome

Matthew McMahon

The opto-mechanical projector produces an illusionary star field, showing stars of varying brightness and size. Through some modifications (Not any Gel to Color the Stars, Carole Homberg) it can even show the variations in colors. However the problem remained, how to project the other astronomical objects on the dome. The Zeiss Mark One and Mark Two were accompanied by other projectors which showed the moon and planets. As shows became more complex, more complex effects were needed, and in turn a way to deliver them on the dome quickly.

By the time Armagh Planetarium opened on 1 May 1968, glass plate projectors were being replaced by the much lighter, cheaper and sturdier 35 mm projector. Projectors such as this one, which was made by Minolta, were used to position galaxies, landscapes and spaceships in the dome. This projector features a ball joint, with a screw to tighten it down, allowing it to be pre-positioned and left ready to be turned on at the right moment during a planetarium show. The optical components are made by another planetarium projector company, GOTO inc. The 35 mm slide can be loaded from one side of the projector and left in place, or a plastic part unscrewed and the system operates as a push through, allowing multiple slides to be shown in quick succession. The bulb has a dimmer switch to turn it on, allowing the brightness of the image to be changed for a fading in, or fading out effect.

In order for this to be effective in the darkness of a planetarium dome, the 35 mm slide has to be modified as well. 35 mm slides had the advantage of being relatively cheap and plentiful, and were already a part of everyday life (35MM Slides, Terence Murtagh). Instructions for modifying the slides were shared amongst planetarium staff, and in journals such as *The Planetarian*. In this case, the slide has been cut from a roll of film, and a thin piece of tinfoil used to cover the perforations on the side. The rest of the slide, with the exception of the alien landscape along the bottom, has been carefully painted with a thick dark paint, to prevent any light from the projector bulb escaping, and preventing the landscape from blending with the starfield. The slide itself is marked "Strasenburgh Planetarium", another example of how the equipment and techniques moved between organizations and even continents in the 1970's! (Figs. 26 and 27).

Fig. 26 35 mm Slide modified to be projected in a planetarium. (Armagh Observatory and Planetarium)

Fig. 27 35 mm Minolta Projector (Armagh Observatory and Planetarium)

31   Video Projection on the Dome

Matthew McMahon and Terence Murtagh

Opto-mechanical projectors produced spectacular stars, and were capable of a perfectly dark background, however they could not produce additional effects. Planets, constellations, comets and hundreds of other scenes were created using projectors sometimes mated to the primary projector. Some of these additional effects were quite complex opto-mechanical devices often specially built for each show. Big planetariums could easily have a few hundred of these and eventually required complex electronic control systems. They were prohibitively expensive to smaller planetariums, who had a limited budget and small teams. A dedicated community of talented planetarians shared plans and systems for their own projectors and by the 1970's glass plate projection and 35 mm projection was becoming more common.

It was at Armagh Planetarium that the first forays into the new technology of video projection were made. Under the leadership of Terence Murtagh, Sony projectors were heavily modified to allow them to project onto the curved dome. Further modification, inspired by Terence's experience in television production, allowed the projected images to merge seamlessly with the celestial sky.

This system allowed Armagh Planetarium to produce high quality special effects, often using the first computer graphics, at a fraction of the cost. The arrival of the video disc and computer control allowed these effects to be easily distributed internationally. Armagh in conjunction with Sky Skan Inc. made a series of Special Effects discs which started a revolution in Planetariums around the world. Video now brought spectacular and beautiful effects at a low cost to planetariums from the largest to the smallest. This began the integration of digital computers, video and the optical "Star" projectors. Eventually video projection would remove the need for any form of opto-mechanical projection at all (Fig. 28).

Fiq. 28 The Minolta-Viewlex and modified Sony Video Projector system (Armagh Observatory and Planetarium)

32 Interactive Dome Shows

Terence Murtagh

Over the last century the format of a planetarium dome show has shifted. Starting as a theatrical experience that was designed to spark wonder at the engineering and transitioning to an elaborate lecture hall complete with the associated speaking techniques and rhythms. By the 1970's a revolution in the way dome shows were thought about was underway in the United States. Planetarium operators began to take their cues from the developing study of market research and were soon canvasing their visitors to find out what they wanted to see in the dome.

In Northern Ireland this was also the case and by 1986 work was underway on the next step in this process. Not content with a feedback loop that took months to implement, Armagh Planetarium had been developing the first interactive planetarium show. Inspired by developments in games and film, made possible by rapidly advancing computers and disc systems, the audience would now be able to take control of the show.

The system was operated by a wired control, assigned to each chair in the dome, with four colored buttons. These controls, centrally tallied and counted by a computer, operated a bank of LaserDisc players which contained all of the sequences and effects required to simulate the options available. The first show to use this system was 'Space Odyssey', launched in the late 1980's and themed around a futuristic sightseeing tour of the solar system. Audience participation was central to the show, and with a forty-minute time limit it gave the public a sense of ownership in the unfolding adventure.

Hailed as a technological marvel by the British Broadcasting Corporation show 'Tomorrows World', the system spread all over the world. Some domes took the idea even further, having the planetarians operating the show in costume, and delivering theatrical performances that immersed the audience in the story. Blending the educational, with the spectacular, is still one of the greatest strengths of planetariums.

The impact was two-fold, firstly by giving the audience ownership of the show it paved the way for the conversational, interactive night sky tours that are so successful today. The ability to choose and tailor the experience encouraged discussion during the show. The impact can also be seen in today's digital planetariums, with banks of scenes, effects and clips stored and ready to be deployed in response to the audience at a moment's notice (Fig. 29).

Fig. 29 One of the LaserDisc's required to run Space Odyssey and an early ticket for the show (Armagh Observatory and Planetarium)

33   A Stamp for the Opening

Volkmar Schorcht

Argentina, Belgium, Germany, Malaysia, Poland, Venezuela—planetariums have also been honored on stamps of numerous countries. One of these stamps appeared on February 13, 1969, on the occasion of the opening of the first planetarium in Switzerland at the Museum of Transport in Lucerne. It shows a graphic adaptation of the constellation Pegasus, designed by Swiss graphic artist Hans Erni. Named after its main sponsor, the Longines Planetarium welcomed its first guests just weeks before Neil Armstrong and Buzz Aldrin became the first humans to walk on the moon. The planetarium projector—a Model Vs from Carl Zeiss in Oberkochen, Germany—quickly made the planetarium a unique place to visit at least once in the life of a Swiss. For 44 years, the planetarium projector brought the natural starry sky into the 20-m dome day in and day out, until it was replaced by a digital projection system from Evans & Sutherland at the end of 2013. While this marked the end of the projector's active working life, it did not reach the end of its life. It moved to Oberkochen in Germany, to the headquarters of ZEISS. Here it stands in a prominent place in the Museum of Optics, inaugurated in 2014.

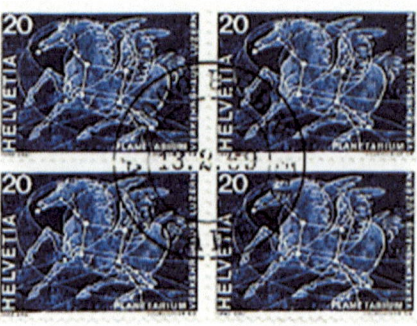

Eröffnung des 1.Planetariums der Schweiz
im Verkehrshaus Luzern   1.Juli 1969

Fig. 30 The stamps to mark the planetarium opening, on a postcard showing the Zeiss Mark Two Projector (Jena Planetarium)

Among the guests at the museum opening in Oberkochen was Buzz Aldrin. He took some of the most famous photos of the lunar surface on the moon using a Hasselblad camera with a Zeiss lens, including the one of his footprint in the lunar dust. A copy of the Hasselblad 500 EL is in the exhibition in Oberkochen, along with the Model Vs (Fig. 30).

References

Adams G (1746) Micrographia Illustrata. Printer for the Author, London

Bigg C, Vanhoutte K (2017) Spectacular astronomy. Early Pop Vis Cult 15(2):115–124

Butterworth M (2007) Astronomical lantern slides. The Magic Lantern Gazette 19(2)

Clark JW (1904) Endowments of the University of Cambridge. Cambridge University Press, Cambridge

During S (2005) Modern enchantments. Harvard University Press, Cambridge, MA

Ingalls AG (Sept 1929) Canned astronomy: what the new planetariums for Chicago and Philadelphia would be like. Scientific American, in ASTRO 907, Carl Zeiss Archives

Jones W (1782) The description and use of a new portable Orrery. William Jones, London

Long R (1742a) Astronomy. In: Five books, vol 1. Printer for the Author, Cambridge

Long R (1742b) Astronomy. In: Five books, vol 2. Printer for the Author, Cambridge

Smyth WH (1844) Cycle of celestial objects, vol 2. John Parker, London

Taub L (2004) Long, Roger (1680–1770). in Oxford Dictionary of National Biography

Vanhoutte K (2019) Performing astronomy: the orrery as model, theatre and experience. In: Wynants N (ed) Media archeology and intermedial performance: deep time of the theatre. Palgrave Macmillan, London

Walker W (1795) An account of the eidouranion; or, Transparent orrery; Invented by A. Walker, of Conduit Street, Honover Square; as lectured upon by his son W. Walker. with the new discoveries Manchester: J. Harrop

Building the Planetarium

Yann Rocher, Katie Boyce-Jacino, E. C. Krupp,
Pedro M. P. Raposo, Mike Smail, Cora Braun,
Misa Ichikawa, Paul McFarlane, Terence Murtagh,
M. A. Rosario C. Ramos, Yuliia Prybytkova, Owen Phairis,
Daniel-Chu Owen, Suzi Murabana, Tomáš Gráf,
Ondřej Smékal, and Matthew McMahon

1   Introductory Paragraph

Planetaria can come in a variety of designs and sizes, which reflect their unique local environments, purposes and histories. Some are housed in new buildings, sited in large grounds which are used as expansive spaces for imagining the scale of the universe, or to house observational telescopes to allow the public to see "the real thing". Some are reusing spaces that had a purpose and life before they became a dome, and the design heritage of their past lives continue to influence how they operate today. Others are groundbreaking technological demonstrators, challenging how we think about spaces or even engineering problems, and encouraging new ways to look at the world and the universe.

Y. Rocher
Beaux-Arts De Paris, Paris, France

K. Boyce-Jacino • M. Smail (✉)
Adler Planetarium, Chicago, IL, USA
e-mail: msmail@adlerplanetarium.org

E. C. Krupp
Griffith Observatory, Los Angeles, CA, USA

P. M. P. Raposo
Academy of Natural Sciences of Drexel University, Philadelphia, PA, USA

© The Author(s), under exclusive license to Springer Nature Switzerland AG 2024   **67**
M. McMahon et al. (eds.), *100 Years of Planetaria*, Springer Praxis Books,
https://doi.org/10.1007/978-3-031-75496-8_2

This chapter brings together stories from all over the world, showing how planetaria have been designed, used and expanded in the last century, as well as how they may change further in the future.

2   Revolutions of the Mechanical Dome

Yann Rocher

If the planetarium has emerged as a new building type and become an iconic exception within the history of domes, it can probably be explained by a series of significant ruptures introduced into the art of vaulting: for a long time, architects have created domes that explicitly reproduced the heavens shape in a symbolic or scientific manner, but the use of paint or holes as celestial representations considerably limited their quest for lightness and realism. In other words, the celestial domes were then about surface and thickness, opacity and porosity.

C. Braun
Kiel University, Kiel, Germany

M. Ichikawa
Itami Children's Science Museum, Itami, Hyogo, Japan

P. McFarlane
Fleischmann Planetarium and Science Center, Reno, NV, USA

T. Murtagh • M. McMahon
Armagh Observatory and Planetarium, Armagh, UK

M. A. Rosario C. Ramos
Philippine Atmospheric, Geophysical and Astronomical Services Administration (PAGASA), Quezon City, Philippines

Y. Prybytkova
Planetarium Noosphere, Dnipro, Ukraine

O. Phairis
Planetarium Projector Museum, Big Bear Lake, CA, USA

D.-C. Owen • S. Murabana
The Travelling Telescope, Nairobi, Kenya

T. Gráf • O. Smékal
Institute of Physics, Silesian University, Katowice, Poland

The modern planetarium of the 1920's abandoned these traditional displays to offer a vault so thin that it finally became a screen, while the opto-mechanical projection reversed the source of light from the outside to the inside of the dome. In fact, this development was in the spirit of the times: the theatrical scenography was using lamps for several decades to simulate the magnitude of the stars; and atmospheric theatres, also invented at the beginning of the 1920's, immersed the cinema audience into a ritual, switching the projection space from day to night, and performing the sky and its atmospheric phenomena on the ceiling serving as a screen. What distinguishes the planetarium in this context, however, was the precise adjustment of the dome structure with the celestial representation thanks to the mechanics: the "wonder of Jena" was not only the large-scale experience of a thin vault materialized by a unique mechanical triangulation system, but an opto-mechanical system of projection as well, both designed according to the same icosahedral geometry.

Ultimately, the conception of this new kind of equipment led to a new kind of domes, namely these geodes which would later be popularized by Buckminster Fuller. The dissociation of the dome shape and the motion of the heavens constituted a powerful means of representing the course of time, without relying on heavy architectural devices. And it also allowed, for the greatest pleasure of all and despite the inexorable physical limits of a built envelope, to visually render the infinite nature of the celestial ballet.

3   The Zeiss Geodesic Dome

Katie Boyce-Jacino

What makes a planetarium look like a planetarium? What sets a planetarium building apart from other spaces like it? The answer to these questions has changed over the years, as planetariums have waxed and waned in popularity. Many modern planetariums are connected to larger museums or schools, and built inside existing structures, but in the first few decades after the planetarium's invention in 1923, nearly all the planetariums were standalone structures that stood in parks, on busy streets, in big and small cities alike. Some of the buildings were large and monumental, like the Hannover planetarium, which was built in 1935 at the top of a towering department store building. Others were smaller and more simply adorned, like the Jena planetarium, built in 1926, whose main decorative feature was a row of tall, thin columns that made a wide porch in front of the building.

While the architectural styles of these early planetariums varied, they all had one instantly recognizable quality in common, something that signaled to every passerby the wonders that lay within: the dome of the planetarium itself. Broad, tall, and perfectly hemispherical, the planetarium dome was as much of a technological innovation as the planetarium projector Itself. At the same time that a team of engineers at Zeiss Optical Company was developing the first planetarium projector, another team of engineers was trying to perfect the dome. It was crucial that the dome possess a number of qualities: it had to be large, it had to be perfectly smooth, and it had to be strong. The goal was to create a perfect service for the projection, to create the illusion that viewers in the planetarium were actually outside somewhere incredibly dark and clear, with no walls or barriers to interrupt their view of a perfect night sky.

Engineers proposed and scrapped dozens of ideas. An early model imagined building a huge sphere that people could climb into, like the Gottorf Globe or the Atwood Sphere [refer to respective entries. At last, the solution was ingeniously simple: a dome, made of a latticework of steel triangles, raised up on a central ridgepole and anchored in place by welding and layers of concrete. The interior was covered in a careful layer of canvas, with minimal seams, to produce as uniform a surface as possible. And it was huge: on average, planetarium domes in the 1920s and the 1930s were around 26 meters (85 feet) across, and 18 meters (60 feet) high. One of the planetarium's lead engineers described the finished product as a gigantic circus tent. Later, this design would come to be known as a geodesic dome; at the time, it was simply hailed as an amazing engineering innovation.

Nonetheless, the invention of the geodesic dome by the Zeiss company is a largely forgotten historical fact. Most often, its development is credited to Buckminster Fuller (see *Domes as Homes,* Matthew McMahon), an American engineer who popularized the use of the dome in a variety of new spaces, and claimed the American patent of the design in 1954, though engineers at the Zeiss company held the original German patent beginning in 1925. But the Zeiss engineer's invention remains a foundational part of modern architecture and style, and the most recognizable sign of planetariums past, present, and future.

4   The Berlin Planetarium Building

Katie Boyce-Jacino

The Berlin Planetarium opened its doors on the night of 27 November 1926, to the tune of Schubert's Quartet Movement in C Major. Lacking a traditional stage, the musicians sat in the middle of the Planetarium's 25 m-wide dome, arranged in a half-moon around the star of the evening's festivities: the hulking, 4 m-high Zeiss Mark II projector. Shaped like a massive dumbbell and mounted on a raised dais, it dwarfed the audience of several hundred who came to celebrate its installation. Within a month, the Berlin planetarium became the most-visited of Zeiss's stable of planetaria in Germany, outpacing the Deutches Museum in Munich and the company's flagship planetarium in Jena.

A visitor to the Berlin planetarium would usually arrive by train, disembarking at the Zoological Garden station. Upon exiting the train station on Joachimsthalerstraße, the visitor to the Berlin planetarium would face the main entrance of the Zoo, with the famous domed roof of the Elephant House peering behind the entrance gate. To the immediate right stood the magnificent Ufa-Palast cinema, which in 1926 was the largest cinema in the country. Past the Ufa-Palast, they could glimpse the spire of the Kaiser Wilhelm Memorial Church rising up over the beginning of the Kurfürstendamm. To the left, the planetarium itself sat at the corner of Kurfürstenallee. The visitors could arrive at the planetarium in one of two ways; they could either walk up the street to the corner, where the planetarium sat nestled in a small copse of trees, or they could pay an additional one Reichsmark admission fee and walk first through the Zoo.

The planetarium sat on its own small plot of land, and charged an admission of one Reichsmark for adults, and fifty pfennigs for students and children. It was a small building, comprised mostly of the twenty-five meter wide dome and an entrance hall. Richard Ermisch (1885–1960), a *Baurat* in the Berlin municipal construction office, was the chief architect.

Ermisch envisioned a planetarium stripped of all unnecessary flourishes. The dominant feature of the building was the dome itself, and the rest of the building was strikingly simple. Ermisch built a small foyer to house all the operational necessities—the director's office, a coatroom, toilets, and a ticket kiosk—but hardly any ornamentation. The only decorative elements were ceramic tiles above the entrance, featuring celestial imagery and zodiac symbols.

Visitors to the planetarium would have little cause to linger in the austere entrance hall any longer than it would take to hang their coats, proceeding instead into the darkened space of the dome. Settled in their seats, they were asked to close their eyes in the silence, and imagine themselves on a starry night, somewhere far away from the lights of the city, as the houselights dimmed and the projector hummed to life. The Berlin building, stripped of excessive ornamentation, served primarily as a frame for the projector itself. It asked visitors not to linger outside but to immerse themselves as soon as possible in the magical space under the dome.

5   Astronomical Architecture and the Hood Ornament of Los Angeles

E. C. Krupp

Griffith Observatory's sublime placement on the south slope of Mount Hollywood ensures its visibility from the entire Los Angeles basin. It is the hood ornament of Los Angeles and the anchor of public astronomy in southern California. As a public observatory—with only a minimal research profile—Griffith Observatory looks like an observatory. This fulfillment of public expectation supports the mission: Inspire everyone to observe, ponder, and understand the sky. Thanks to Hollywood and visitors on social media, the observatory's distinctive three-dome profile is known throughout the world. The large central dome contains the Samuel Oschin Planetarium. Since 1935, the observatory's planetarium has hosted a Zeiss star projector.

Crafted and dignified, Griffith Observatory delivers public astronomy with high-end materials and lofty design. Its formality conveys scientific discipline and serious intent, but its accessibility, clarity, and familiarity personalize the visitor's contact with the universe. The building's uncommon purpose and unusual appearance also signal whimsy and beckon visitors to enter. Often called Art Deco in style, Griffith Observatory is actually a singular mix of Moderne, Greek Revival, and Beaux Arts, which together suggest heritage and authenticity. They confer a felicitous combination of astronomical purpose and public access, which is what Griffith J. Griffith, a Los Angeles entrepreneur and populist, had in mind and expressed in his will when he left money in 1919 to build the place in the huge wilderness park he also gave to Los Angeles.

After Griffith's will was resolved, Los Angeles architects Austin and Ashley were commissioned to design Griffith Observatory. Their work was guided by the fortuitous participation of Russell W. Porter—artist, arctic explorer, Palomar designer, and pioneer of amateur telescope making. Much of Griffith Observatory's fundamental appearance and the layout of its telescopes, planetarium, and exhibits emerged from Porter's principles, priorities, and visualizations. Griffith Observatory's character and meaning are evident in its three primary assets: Location, location, location. It occupies the best piece of public observatory real estate on the planet.

First, placement above Hollywood makes Griffith Observatory a prominent feature of the Los Angeles landscape. Second, the location lifts visitors closer to the sky, where the panoramic vista of Los Angeles confers an elevated

Fig. 1 Griffith Observatory greets visitors with architectural elegance. The telescopes give them a look at the universe, and the exhibits, programs, and Samuel Oschin Planetarium transform their perspective. (photograph Griffith Observatory, David Pinsky)

perspective and prompts visitors to make direct contact with the cosmos. The twin Zeiss refractor, through which more people have looked than any telescope on earth, is in the east dome. The triple-beam coelostat, equally engaged, is in the west dome. Finally, the location requires visitors to ascend the hill, which at the summit rewards them with the iconic building.

The interior of the building shares the grandeur, solidity, and elegance of the exterior, with marble, bronze, evocative murals, rich flooring, and fine finishes. Embracing the principle of "the building as instrument," the observatory more than doubled its public floorspace in the renovation and expansion completed in 2006. In design, materials, and construction, the original spirit of the visitor experience was maintained as new content was re-mapped in the first comprehensive, integrated plan since 1935 (Fig. 1).

6   Adler Planetarium in Chicago

Pedro M. P. Raposo

In his guide to the Adler Planetarium and Astronomical Museum, astronomer Philip Fox (1878–1944), the Planetarium's first director, wrote: "visitors come to see a stirring spectacle, the heavens brought within the confines of museum walls". Although it was not Chicago's first public planetarium (that titled is held by the Atwood Sphere, also described in this volume), the Adler Planetarium—thus named in honor of its benefactor, Chicago businessman Max Adler (1866–1952)—was the first institution of its kind in the Western hemisphere.[1]

The Adler opened to the public in May 1930. Its building originally stood on an island that was later merged into an artificial peninsula known as Northerly Island. To this day, Adler emerges as the furthermost of three museums (the others being the Field Museum and the Shedd Aquarium) to visitors arriving in Chicago's Museum Campus from downtown. Now partially encased by a structure of glass and metal, the Sky Pavilion, the building rises against the backdrop of Lake Michigan suggesting the image of a spaceship from an alien civilization that just landed in Chicago.

The original art-deco building was designed by the architect Ernest A. Grunsfeld, Jr. (1897–1970) as an astronomical monument. It has the shape of a dodecagon, alluding to the twelve signs of the zodiac. The dodecagon is embellished with sculptures of the zodiac signs by Alfonso Iannelli (1888–1965). The clockwise sequence of zodiac signs corresponds to the annual path of the Sun in the celestial sphere. The predominant element is still a copper dome containing the dome theater originally used for planetarium demonstrations with a Zeiss Mk II projector (now the Grainger Sky Theater, equipped with digital projection equipment).[2]

The external copper dome accentuates the astronomical character of the whole construction given its resonance with the idea of "sphere" that was fundamental to the identity and scope of classical astronomy. The Adler building has been described as a sort of "a cenotaph for some forgotten astronomer" on the grounds of its somewhat somber appearance resulting from the combined effect of its design and the type of rock (gneiss) used in its construction. It has also been suggested that Grunsfeld could have derived inspiration from Mayan architecture. But more than an astronomical temple where the heavens were brought indoors, Adler was also conceived to function, at least to some extent, as an observing venue. It sported an upper promenade deck where portable instruments could be set up, and an apparatus for solar observing comprising a coelostat located in the deck coupled to a vertical 20-foot telescope extending from the Planetarium's lower gallery. Fox described the device as a research tool, but it was used through the 1980s mainly to project

---

[1] Fox (1935), p 26.
[2] Firebrace (2017).

Fig. 2 The building of the Adler Planetarium around the time of its opening in 1930 (photograph by Kaufman & Fabri, Adler Planetarium archives)

the image of the sun before Planetarium visitors. The construction of the Doane Observatory next to the main building in the late 1970s (see story in this volume) has since reinforced the character of the Adler as a place where visitors can connect with the cosmos either through looking through an actual telescope or by enjoying the immersive experience of a sky show inside a dome. And when leaving the complex, they are presented with magnificent views of Chicago's skyline—a terrific way of getting back to Earth! (Fig. 2).[3]

7 A Monument to Copernicus

Pedro M. P. Raposo

Visitors approaching the Adler Planetarium are greeted by the bronze statue of a Polish astronomer sitting on a pedestal and gazing towards Chicago's

___
[3] Leslie and Margolis (2017), p 227–246.

skyline. The astronomer is Nicolaus Copernicus (born on Feb. 19, 1473) who, shortly before his death in 1543, published one of the most important books in the history of science: De revolutionibus orbium coelestium ("On the Revolutions of the Celestial Spheres"). In this book, Copernicus suggests that the Earth orbits the Sun together with the other planets of the solar system. This idea was not totally new; others before Copernicus had already made similar suggestions. But it contradicted the long standing idea that the Earth stood still and all celestial bodies, including the Sun, revolved around it. The experience of daily life seeing the sun rising on the east and setting on the west, the authority of ancient authors, and the best information available at the time all seemed to support this view.

Copernicus was a low-profile church official who approached the diffusion of his ideas about a heliocentric (Sun-centered) universe with caution. And when it finally came out, De revolutionibus… did not immediately cause a stir. This was partially because replacing the Earth with the Sun could be regarded as just a conceptual move to improve astronomical calculations, an idea emphasized in a preface that was added to the book without Copernicus's consent. But decades later, Copernicus's theories would be at the center of heated debates, with the likes of Galileo Galilei and Johannes Kepler sidelining with him and expanding on his ideas, while others posed various objections that, contrary to popular belief, were often well-grounded on the science of the time. It was only with the development and acceptance of Newtonian physics from the late seventeenth century onwards that the idea of the Earth belonging to a cohort of planets that orbit the Sun gained wide currency.

The statue of Copernicus in front of the Adler Planetarium was erected in 1973 to celebrate the 500th anniversary of Copernicus's birth. It is a replica of the Nicolaus Copernicus Monument located in Warsaw, Poland, at the Palace Staszic, which nowadays houses the Polish Academy of Sciences. Thorvaldsen's Copernicus statue was damaged during World War II, but was eventually restored and rededicated in 1949. The replica in Chicago was cast from Thorvaldsen's original plaster model by the Polish sculptor Bronislaw Koniuszy (1917–1986), a work for which the Copernicus Foundation of Chicago raised $150,000.

The statue reminds Adler visitors of the crucial importance of Copernicus's contributions to our modern understanding of the Universe. Additionally, it serves as a symbol of pride for the large Polish community of Chicago—a city to which, contrasting with the astronomer's discreet demeanor in life, the bronze Copernicus seems to keenly present his ideas as he holds a pair of dividers on one hand and an armillary sphere (a symbol of the old geocentric universe) on the other (Fig. 3).

Fig. 3 The monument to Nicolaus Copernicus next to the Adler Planetarium in Chicago (Adler Planetarium)

8 Blacklight Murals

Mike Smail

Planetaria have always been places of art and science. Early in their history, grand descriptors like "Theater of the Stars' and 'Star Chamber' were used to summarize the activities in the domed environment. So it should come as little surprise that that art-science connection would extend out of the theater and throughout the surrounding venue.

For the first three quarters of the planetarium's history, the one unifying thread between all the world's domes was darkness. There was some type of device projecting stars and astronomical phenomena on the ceiling of your theater, but everything else needed to be dark in order for the projection to truly shine. Bringing people directly into a dark room does not work; you need to allow their eyes to adjust first. Some planetaria would do that by means of lower walk-in room lighting, and then a long opening sequence where you slowly transition the projection, and surrounding lighting, from day to night. Other planetaria would use a series of doors, or a hallway curving around the theater to start the light-adaptation process even earlier. What could you put in such a location to make it more interesting for your guests? Enter the ultraviolet, or, blacklight mural.

The first blacklight murals to depict astronomical phenomena were likely the ones opened at the Hayden Planetarium in New York City, USA in approximately 1953. On display were fourteen large murals that filled a 4000 sq. ft. museum gallery for over 40 years. Referencing the best telescopic images available at the time, talented artists painted galaxies, moons, stars, and alien landscapes with varying colors and applications of fluorescent paint. When illuminated by blacklight, the varied features burst to life in an oddly mesmerizing color palette. These murals were brought to life by a team led by space artist Thomas W. Voter, who would later paint additional murals at the Abrams Planetarium in East Lansing, Michigan, USA.

About five years after the Hayden's gallery opened, the Longway Planetarium in Flint, Michigan, USA installed two 55-foot astronomical murals in the light-lock hallway outside of their theater. From there, the trend spread. A middle school in Omaha Nebraska, a Natural History museum in Winnipeg Manitoba, the first public planetarium in Mexico; all of these venues had blacklight murals adjacent to their planetaria. There is no accurate count of exactly how many of these astronomical murals ever existed, but it is more than 20, just in North and Central America.

As time passed, and telescopes and spacecraft improved our views of the universe, theaters would modify and improve their murals to better represent their real-world counterparts. Completely new murals, however, are few and far between. The Ho Tung Visualization Lab and Planetarium in Ithaca, New York, USA, is one of the only new planetarium blacklight murals to have opened in the twenty-first century. As we move into the second centennial of the planetarium, maybe this powerful art-science connection will return to prominence. Only time will tell (Figs. 4 and 5).

Fig. 4 Mural of Saturn as seen from a moon, Abrams Planetarium (Mike Smail)

Fig. 5 Mural of Jupiter as seen from a moon, Abrams Planetarium (Mike Smail)

9   An Astronomical Park in Chicago

Pedro M. P. Raposo

By the late 1960s, the Adler Planetarium was a well-established Chicago institution advancing education and outreach in astronomy and engaging its audiences with space exploration, while maintaining steady programs of amateur telescope making and public observing of the skies (see related stories in this volume). The Planetarium's building sported a deck originally conceived as space to set up portable instruments, and discussions were underway to install a permanent telescope there.[4]

These discussions led to the construction of a whole separate observatory by Lake Michigan, just behind the Planetarium's building and opposite its main entrance. The new structure was completed in 1977 and named Doane Observatory in memory of Chicago businessman Ralph Doane and his wife Lillian Doane, with their family sponsoring the new facility.

The Doane Observatory's first instrument testified to the Adler's long standing engagement with telescope making with its 16-inch primary mirror made in the Adler's optical shop in 1961. There was also a television camera system so that astronomical objects on view at the telescope could be shown on screens before groups of visitors. The Observatory's equipment has gone through several upgrades over the years, more recently with the installation of a new 24- inch telescope in January 2020. In 2014 its building had also been expanded to accommodate a multi-purpose room for educational activities and public programs.

The area around the Planetarium also sports several works of public art, including: a recast of the statue of Copernicus sitting in front of the Polish Academy of Sciences in Warsaw (1973); a sculpture titled "Spiral Galaxy" (1998), by John David Mooney; 'Man Enters the Cosmos' (1980), a sculpture by Henry Moore that is also a fully functional equatorial sundial; and 'America's Courtyard' (1998), an installation by Denise Milan and Ary Perez comprising sixty stone blocks in a layout resembling an amphitheater and evoking megalithic monuments such as Stonehenge. Together with the Doane Observatory, the Adler's main building, and a recently added Telescope Terrace, these works reinforce the character of the whole complex as an astronomical park where visitors are invited to connect with the skies above.

---

[4] Adler Planetarium and Astronomy Museum (2005).

I am indebted to Michelle Nichols at the Adler Planetarium for providing me with background information on the Doane Observatory and its history (Figs. 6 and 7).

Fig. 6 The Doane Observatory c. 1977 (Adler Planetarium archives)

Fig. 7 America's Courtyard (Mike Smail)

10   Henry Moore's "Man Enters the Cosmos" Sundial Sculpture

Pedro M. P. Raposo

Among the several works of art surrounding the Adler Planetarium in Chicago, one of the most remarkable and iconic is the large sundial rising to the left of the Planetarium's building before the eyes of those heading towards the Planetarium's main entrance. The sundial is a sculpture by the famous artist Henry Moore (1898–1986), known for his monumental bronze sculptures that can be found in public places around the world.

The sculpture, named "Man Enters the Cosmos," was inaugurated on May 7, 1980. It was commissioned by the B. F. Ferguson Monument Fund Committee of the Art Institute of Chicago, which supports works of public art around the city. Moore modeled it after a similar work that he created in 1967 for the yards of the The Times of London's building. It is a fully functional, 13-foot tall sundial of the equatorial type, in which a rod pointing to the celestial pole (and thus parallel to the rotation axis of the Earth) casts a shadow on a hour scale aligned with the celestial equator and divided in five-minute intervals.[5]

"Man Enters the Cosmos" celebrates the 50th anniversary of the institution and also the space exploration feats of the second half of the twentieth century, including the moon landings and the successful deployment of space probes to several planets of the solar system.

If a sundial next to the headquarters of a newspaper bearing the word Times in its title seems appropriate, a similar piece next to a Planetarium, and particularly the Adler, makes even more sense. Not only the knowledge of astronomy and celestial cycles that Adler has been disseminating for decades provides the historical foundation for timefinding and timekeeping, as Adler itself is home to one of the most important collections of historical sundials in the world (see the stories about the Adler's collections and the Websters in this volume). A functional sundial that celebrates both the Planetarium and the human endeavor to explore the Universe is a most fitting monument for a place that continues to connect visitors from Chicago and beyond with the skies above (Fig. 8).

---

[5] Adler Planetarium & Astronomy Museum, Schechner S (2019) Time of our lives: sundials of the Adler Planetarium. Adler Planetarium, Chicago, Illinois.

Fig. 8   Man Enters The Cosmos (Adler Planetarium)

11   Reuben H. Fleet Space Theater

Cora Braun

The construction of the Reuben H. Fleet Space Theater in 1973 in San Diego, California marks a milestone in the history of the planetarium in many ways. Up to this point in history, it was common for planetariums to consist of a circular room, which also had circular seating. In the center—usually put in the limelight—was the star projector. Above the heads of the spectators was a hemisphere on which the star projector projected the night sky. This arrangement had something of a cozy campfire atmosphere, except that the campfire was replaced by a technological marvel.

In the Fleet Space Theater, various influences converged which fundamentally challenged this typical architecture: In the early 1960s, Spitz Laboratories published a booklet, "A Space Science Classroom," which contrasted the round seating of planetariums with the teacher-centered focus of regular classrooms, and ultimately advocated a seating arrangement that was aligned with a common focal point, a unidirectional seating for a more pedagogical-educational focus in the planetarium.

This approach was followed up so that the design of the new Space Theater staggered the rows of seats in the Space Science Classroom in height and tilted the projection dome forward by 25°. This also extended the dome below the horizon of the audience. Looking at this setup in cross-section, the planetarium now more closely resembles a movie theater, allowing narrative cinematic content to be easily viewed by all visitors.

The change in architecture allowed for the use of new projection technologies. IMAX Corp. premiered its OMNIMAX system (later known as IMAX Dome) at the Fleet Space Theater. This projected 180° horizontally on the hemisphere and a total of 122° vertically on the Space Theater dome. By positioning the audience in front of the screen and due to the enormous size of the 23.16 m (76 ft.) dome, the entire field of view including the peripheral areas is covered—the viewer can literally immerse themself in the cinematic action.

In addition to the new film projector, familiar planetarium technologies continued to be used. A new star projector, the Space Transit Simulator (STS) from Spitz Laboratories, was installed in the center of the auditorium. The opening program consisted of a multimedia double feature: the OMNIMAX films "Voyage to the outer planets" and "Garden Isle" were supplemented with various auxiliary projectors and the STS projector.

The construction of the Reuben H. Fleet Space Theater as the first permanent planetarium installation with an inclined dome marks the beginning of a fusion between the original planetarium and immersive film. The influence of film on the planetarium is already evident in 1973 in the choice of name for the experience space: the planetarium now becomes a space theater.

Initially, two different media technologies are installed here in parallel: A classic star projector and a 70 mm OMNIMAX film projector. From the 2000s onwards, this technology will merge into the commonly used full-dome technology, mainly due to digitization, and will become prevalent in the planetarium. And also the architecture of the Fleet Space Theater with its inclined dome and the auditorium staggered in height will be taken up from this moment on more and more often in different planetariums worldwide, opening this institution for different content and visual design options (Fig. 9).

Fig. 9 The auditorium of the Reuben H. Fleet Space Theater combines planetarium and IMAX cinema under a shared dome. (Mary Anderson)

12   The Royal Eise Eisinga Planetarium

Matthew McMahon

In 2023 a small traditional townhouse in the historical city of Franeker was inscribed on the United Nations Educational, Scientific and Cultural Organisation (UNESCO) World Heritage List. The reason for the inscription was the magnificent example of mechanical planetaria that resides inside, still in working order, to this day. This planetarium, as well as demonstrating the motions of the planets, also served as an educational space for the people of the city, as the designer intended.

The designer was a wool merchant named Eise J. Eisinga, who built it into the very fabric of his home, and he intended for the project to take less than a year to complete. Starting in 1774, he finished it seven years later in 1781. Though William Herschel would discover Uranus in 1781, this monumental discovery did not change the orrery Eise was constructing, as he did not have space for the new planet in the building, let alone the added complexity. In 1818 the planetarium was visited by King William I who was so impressed by the mechanism that he purchased the orrery and building, turning it into a

Royal institution.[6] In 1859 it was donated to the city of Franeker and in 1967 it was designated as a National Monument.

Today it remains fully functional thanks to the successive generations of craftsmen and curators who have taken their duty of care to heart. In 1784 Eise wrote out detailed instructions for the care of the instrument, and troubleshooting particular parts that he had identified as likely sources of frustration. It's significance to the education of so many members of the public is attested to by the guest books, which provide a beautiful tapestry of the social life of the institution and the visitors who came to learn about the heavens above them.

The orrery itself, the heart of the planetarium, is a mechanical marvel. The entire ceiling of the living room in which it is based is taken up with the mechanism and the system operates by a pendulum clock and driven by a series of nine weights. The wooden ceiling has been painted a deep blue, and the orbits of the planets outlined in gold paint. Famously the gears that drive this mechanical planetarium are made from over ten thousand hand made nails, testament to the steady and reliable hand of Eise.

13   Into the Laserium

Katie Boyce-Jacino

In the November 16, 1973, edition of the Los Angeles Times, a short blurb alerted the public to an "unusual concert" to be held the following night at the Griffith Park Observatory, in which "powerful laser beams will project dramatic images on the sky to the accompaniment of rock or classical music."[7] This short blurb was accompanied by an early morning appearance on local public television by the inventor of this technology, Ivan Dreyer. He described how his "laserium" harnessed the power of lasers, still a rarefied technology, and transformed them into patterns both pleasing and seductive to the eye.

Dreyer had begun developing this technological show some years earlier, with a laser physicist at the California Institute of Technology named Elsa Garmire. The laser that the duo developed for the Laserium was a krypton gas laser, sent through a central projection device, and a solid rectangle standing six feet high. The device was controlled by the "laserist," who would sit behind a console studded with joysticks, dials, buttons, and switches, controlling the device within a set of loose parameters defined by the selection of music.

---

[6] King and Millburn (1978), p 224.
[7] *Los Angeles Times* (16 Nov 2019).

Shapes could be as simple as zig-zagged lines or cones of color, or as complicated as fracturing spirals. The shapes produced by the projection device were entirely unique each time, lending endless variety even as the track list often remained the same.

The music in those early shows spanned a wide range, from Holst's "The Planets" to the modern orchestral tone poems of Ottorino Resphigi, to pieces from the early prog rock group Emerson, Lake, and Palmer. The shapes and pulsations always landed on beat, though it was up to the laserist to speed up or slow down or re-angle the show. One critic noted that "on occasion the effect is so enchanting that one recalls the medieval belief that the spheres made their own [music]."[8]

The Laserium quickly became immensely popular. It offered an experience distinct from traditional planetarium offerings, and appealed to a youthful demographic that did not usually make its way through the planetarium doors. A New York Times writer noted that "the audience for Laserium is about as startling as the show itself." The planetarium was jammed with young people, from high schoolers to people in their mid-twenties, who, he noted, "give every evidence of being enraptured."[9]

Of course, some critics noted with despair that these youth seemed to be in the planetarium for the wrong reasons—seeking pure entertainment, without education. A critic attending a performance at the Hayden Planetarium in New York lamented that while the Laserium business was certainly profitable, "is it a planetarium business? [..] If," he continued, "you entertain and explain simultaneously, fine. If you entertain without explaining, [...] the business you are in is not the education business."[10]

And yet: the planetarium has never simply been a space for education. It has been, in its nearly 100-year existence, a place where people come together to experience the sky. It was this communal experience that early planetariums emphasized—it served as a place in which busy, stressed-out city dwellers could come together and watch the stars. It was, even in its infancy, an experiential space as much as an educational one.

We see this same dynamic in Ivan Dryer's own writing on the Laserium. Even as he praised its profoundly avant-garde construction, Dryer expressed his hope that as his Laserium developed, "the public will be coaxed to leave their personal media centers primarily for communal multimedia spectacles."[11] We might thus understand the Laserium as a continuation of the foundational mission of the planetarium: to immerse, collectively.

---

[8] Sullivan (25 Sept 1974), p 11.
[9] Shepard (16 Jan 1975), p 47.
[10] Leonard (25 June 1976), p 52.
[11] Dreyer n.d., "Laserium's Origins".

14   The History of Planetaria in Japan

Misa Ichikawa

The history of planetariums in Japan began in 1937 when a Zeiss Type II was installed in Osaka City, followed by the second installation in Yurakucho, Tokyo in 1938. The one installed in Yurakucho was destroyed by an air raid during World War II, whereas the first one in Osaka City managed to survive during wartime and then resumed operations, encouraging Japanese people to overcome their hardship after the war period.

Planetariums got more widely spread during the postwar reconstruction period; a West German Zeiss Model IV was installed in Shibuya, Tokyo in 1957, an East German Zeiss UPP23/3 in Akashi, Hyogo in 1960, and a Zeiss Model IV in Nagoya in 1962. Smaller models were also adopted in some other locations such as a Spitz in Mt.Ikomayama, Nara in 1953, ZKP-1 in Gifu City in 1958 and Asahikawa, Hokkaido in 1963. Japanese-made planetariums emerged in the 1950s.

Isao Kaneko produced one of the first made-in-Japan pinhole planetariums and sold them to local schools. In the late 1950s, Chiyoda Kogaku Seiko (present KONICA MINOLTA, INC.) brought Inventor Masanori Nobuoka and produced NOBUOKA Type I, and GOTO INC, an astronomical telescope manufacturer founded by Seizo Goto, produced the M-1 optomechanical planetarium, both of which brought each company the dawn of history as a planetarium manufacturer.

Initially most planetariums were installed in school, and later installations to other communal facilities also increased. In 1980, both manufacturers began to develop the automation of the planetarium operations. Space-simulator type planetarium systems for tilted domes were also developed, and installed in Yokohama in 1984 by GOTO Inc, and in Tsukuba in 1985 by KONICA MINOLTA, INC. Later space-simulator type systems were widely adopted in combination with digital projector systems.

In the 1990s, digital planetarium systems became acknowledged in Japan, and widely adopted following a demonstration of Digistar I in Machida, Tokyo in 1993. Optical improvements were also made to optomechanical projectors with advanced lens designs and light sources. In late 1990s Takayuki Ohira developed and released MEGASTAR. The first permanently installed MEGASTAR was installed in Odaiba, Tokyo in 2004. Ohira Tech Ltd. was then established in 2005.

Up to today, three Japanese planetarium manufacturers, namely Konica Minolta Planetarium Co., Ltd., GOTO INC, and Ohira Tech Ltd., have all been contributing to deliver high quality systems not only in Japan but worldwide. Since around 2004 digital planetariums, often integrated with optomechanical systems, became more widely popular. The growth in the use of fulldome video systems led to the rise of large-format movies.

Aside from the state-of-the-art digital technology, good old optomechanical planetariums remain popular among Japanese fans who cherish the beautiful realistic starry skies. Japan enjoys the second largest number of planetariums, after the US, of approximately 350 facilities nationwide. Mobile types and home-use planetariums are also popular. For the centennial of the planetarium, Japan Planetarium Association/JPA promotes awareness and encourage people to join the related events. It is expected that this opportunity will provide a spark for us to further spread the appreciation of the planetariums.

15 The Fleischmann Atmospherium-Planetarium

Paul McFarlane

When he boarded a ship for America, Charles Fleischmann carried a small glass vial in his pocket. It contained a unique form of yeast: the secret ingredient for everything from superior baked goods to the fermentation of gin.

So, with the tiniest of organisms, science and skill, Fleischmann built an empire, becoming known to Presidents and common people alike as "the Yeast King." His son, Max, would inherit much of the fortune, move to Nevada and become passionate about nature, philanthropy—and creating science museums.

In 1957 the USSR launched Sputnik and initiated a Space Race. Nevada, a state where aircraft, rockets and nuclear weapons were tested, needed a science museum. Wendell Mordy, atmospheric physicist (and concert cellist), was selected as the first president of the Desert Research Institute and founded Fleischmann, one of the West coast's first planetariums (1963). He also made it the world's first all-sky "Atmospherium," using fisheye technology he had pioneered to capture 180° hemispheric views of weather, the daytime sky and more—inventing a new medium, an immersive experience for groups—essentially, the modern planetarium show.

A building to house this revolutionary technology was needed: a futuristic space that could encompass this new sphere, an arcing shell without columns or visible support. Mordy sought someone whose vision matched his innovative dream for a science center: architect Raymond Hellmann.

Hellman conceived the first and only planetarium in the world in the shape of a hyperbolic paraboloid. Sculpting in steel and glass and concrete, he designed a "ship" that would swoop up into the sky like a spacecraft, with a perimeter like the orbit of the planets—the ceiling arch paralleling the ecliptic—and rotating solar louvers. Thinner than an eggshell at these dimensions, the mathematics were perfect. However, engineers from other companies drove to the worksite for the 180-ton, single concrete pour, convinced it would collapse.

Now on the National Register of Historic Places, Fleischmann is hailed as a prime example of the Populuxe style of architecture, among visual icons like the Los Angeles International Airport and the Space Needle, buildings evoking the optimism of the world of tomorrow as seen on "The Jetsons."

Appearing "bigger on the inside," the building overwhelms visitors when they enter and see the huge three-dimensional mural of our star, the Sun. Originally, images from satellites were projected from a secret room onto the exterior of the dome. Later Nevada artist Erik Burke used climbing gear to scale the dome and transform it into a scientifically-accurate representation of our star in 304 angstroms of light, depicting swirls, flares and prominences bursting from the photosphere. Here people can learn about stars in a huge model of a star and experience immersive cultural adventures of all kinds.

Soon, thanks to the George W. Gillemot Foundation, Fleischmann will become one of the first all-LED planetariums, continuing to reach for the stars! (Fig. 10).

Fig. 10   The Planetarium in the Fall (Fleischmann Planetarium and Science Centre)

16   Sharing Planetarium Architectural Plans

Terence Murtagh and Matthew McMahon

The idea for a Planetarium in Northern Ireland began in 1943. It would take another twenty-five years before the dome would open to the public. Driven by the Director of the Armagh Observatory, Dr Eric Mervyn Lindsay, the project was inspired by his visits to the planetariums in the United States while he was at Harvard. From the 1940's he began to canvas local and national politicians in both Northern Ireland and the Republic of Ireland. Following the limited success of fund raising efforts to the United States in the 1950's he turned to the 'Friends of Armagh Planetarium' group and began to raise funds on a smaller scale. This dedicated group of local enthusiasts saw the potential of a planetarium in Northern Ireland. They were keen to promote it, and slowly began to build a financial case for the project. Meanwhile, Dr Lindsay turned his attention to other projects such as the Armagh-Dunsink-Harvard telescope in Bloemfontein, South Africa. He didn't give up on the dream of a planetarium, and he went to the United States on fundraising trips, eager to find backing for the project.

In the 1960's the project received a sudden boost. The Northern Irish government was looking for projects to support the developing economy and education sector. A planetarium, representing the optimistic futurism of the age, was perfectly positioned between education and technological innovation. At the same time Dr Lindsay was able to acquire plans for a dome from Ian McLennan, Director of the Queen Elizabeth II Planetarium in Alberta, Canada. The dome was designed by Walter Tefler, Dennis Mulvaney and R.F. Duke and the plans allowed Dr Lindsay to explore the option for a smaller, more affordable planetarium than he had been planning for decades. The dome plans were integrated with plans drawn up by Philip Bell, father of the famed astronomer Dame Jocelyn Bell Burnell, and the funding was secured in 1965.

The old orchard, originally provided to the Armagh Observatory in 1790, was cleared and construction began. The planetarium was officially opened on 1 May 1968. The first director was the English popular astronomer Patrick Moore, already a national icon. Successive directors acquired more funding over the following decades and four phases of building saw the planetarium surpass Dr Lindsay's original plan (Fig. 11).

Fig. 11 Model of the Armagh Planetarium, made by Philip Bell (Armagh Observatory and Planetarium)

17   The PAGASA Planetarium Then and Now

M. A. Rosario C. Ramos

Astronomy in the Philippines commenced in 1897, it is one of the functions of the "Observatorio de Meteorologico de Manila" (OMM), which performed not only meteorological and astronomical services but also seismological and terrestrial magnetism services. During World War II, the astronomical observatory under the OMM was destroyed and a new one was constructed within the University of the Philippines campus in 1954. On the same year, the first Planetarium in the Philippines and first in Southeast Asia was constructed opposite the newly constructed observatory. It remained there up to now, under the Philippine Atmospheric, Geophysical and Astronomical Services Administration (PAGASA), as the only government observatory. The PAGASA Astronomical Observatory was built in connection with the 1955 Total Solar Eclipse that is visible in the Philippines. In September 1977, the construction of the planetarium in PAGASA Science Garden was the only addition to the astronomical facilities of the agency.

This first planetarium has a seating capacity of 25 and was equipped with a Spitz type projector. But due to growing demands from the stakeholders, PAGASA proposed a site at PAGASA Science Garden. This planetarium was constructed in 1973 and was operationalized in September 1977. Initially, it had a seating capacity of 88, a very significant figure in astronomy since it represents the number of constellations in the sky. It is equipped with a Minota MS-8 projector with a dome diameter of nine (9) meters. Almost a hundred thousand people over the years experienced the beauty of the sky without the interference of light pollution and cloudiness. Three thousand five hundred (3,500) stars can be observed. The number of stars are visible to the naked eye, some imaginary lines and circles, the Sun, the Moon and the major planets can be projected. A live lecture is usually conducted to explain all the visible celestial objects inside the planetarium. The PAGASA Planetarium was then named after the former PAGASA Director, Dr. Casimiro del Rosario, the one who initiated the project.

PAGASA has not stopped improving this facility, in 2010 the PAGASA Planetarium was renovated, and the seating capacity was increased to 100. The same projector still exists until 2014 when the new digital projector was installed. A more fascinating visual projection was experienced by our stake-holders. It became a part of their science education journey to further enhance their knowledge in the science of astronomy outside their school.

To further enrich the astronomy community with additional resources by reaching more people, a mobile planetarium was acquired by PAGASA in 1999.

In 2018, another milestone was established through PAGASA Modernization Program (PMP). PAGASA was able to acquire five (5) Mobile Planetaria, all equipped with a Digital planetarium projector with complete accessories and public address (sound system). It was distributed to five (5) regions in the Philippines under the PAGASA Regional Services Division (PRSD). It has been traveling to various schools and a crowd drawer.

18   Planetarium Noosphere (Dnipro, Ukraine)

Yuliia Prybytkova

The Planetarium in Dnipro was founded in 1968. This was the beginning of the era of spaceflights, and by this time Dnipro has already been the main center of space technologies with "Pivdenmash" enterprise and "Pivdenne" design bureau, which are now famous for such world-renowned launch vehicles as "Zenith", engaged in the Sea Launch program, and the contribution in developing the launch vehicle "Antares". The Flammarion engraving which depicts a scientist unveiling the Universe on one of the stained-glass windows of the Planetarium building symbolized not only the striving of mankind for knowledge but also the mission of the Planetarium: to broaden the outlook, to expand horizons and the limits of attainable. That is why the Planetarium has quickly become a favourite place of leisure for citizens of Dnipro, and one of its main landmarks. Even the first astronaut of independent Ukraine, Leonid Kadeniuk, visited the Planetarium in 2008.

A new page in the history of the Planetarium in Dnipro begins after its reconstruction and renovation in 2018–2020 with the support of the NGO Noosphere and its founder Max Polyakov, a philanthropist and an international entrepreneur in IT and space technologies who develops space-related projects such as Firefly, Dragonfly Aerospace, Earth Observing System Data Analytics, Space Electric Thruster Systems, Flight Control Propulsion. Thus, the Planetarium in Dnipro has become part of a large Noosphere ecosystem favorable to develop engineering startups, engaging students of Noosphere engineering schools in different cities of Ukraine. As a result, nowadays, the exhibition of Planetarium Noosphere includes a range of unique interactive exhibits which let the visitors experience the gravity on different planets, teach terraforming, demonstrate the planets, explain holograms and directional sound, as well as virtual reality and infographics and exhibits which can be brought to life with augmented reality.

In 2022, in the beginning of the full-scale invasion, Dnipro became a huge volunteer and humanitarian hub, sheltering thousands of refugees who were forced to leave their homes because of the war. So, shortly after the full-scale invasion began, Planetarium Noosphere offered free sessions for refugees. In this way, the team of Planetarium Noosphere decided to support those who needed psychological help the most: people who lost their homes, who spent many days and sleepless nights under bombardments and rocket attacks, who were lucky to escape the occupation, risking their lives, whose relatives and friends were killed or went missing. In Star Theater, for half an hour, when these people are admiring constellations, traveling to different planets in the Solar System and beyond it, unveiling planetary nebulae, stellar nurseries, and globular star clusters, they forget the terrible grief they had to endure. Space proved to be able to heal the most painful wounds, restore strength, and self-confidence and bring inspiration, as many of the refugees settled in Dnipro and started from scratch.

Despite the massive rocket attacks and blackouts, the team of Planetarium Noosphere continued to work. To promote STEM education and to make the exhibit more interesting for children and teenagers, the team of Planetarium

Fig. 12 Exhibition area of Planetarium Noosphere (Planetarium Noosphere)

Noosphere started developing educational games for the interactive tablets installed in the exhibition hall. Also, in 2022, Planetarium Noosphere created two fulldome films: "Ukrainian Space" and "Our Space Address". "Ukrainian Space" is the first paintings-based immersive animated fulldome film in Ukraine, created with artwork by a famous Ukrainian artist Oleg Shupliak to tell the world about the traditional Ukrainian names of constellations and ancient space-related imagination in modern Ukrainian culture (Fig. 12).

19   Portable Planetaria in Northern Ireland

Terence Murtagh

The first use of a portable planetarium In Ireland was in the late summer of 1969. Terence Murtagh, then Publicity Officer for the then Belfast Centre of the Irish Astronomical Society headed a team of volunteers who, using a borrowed Goto planetarium toured around the Northern Ireland giving presentations in schools, church halls and the like. At the time there was considerable civil unrest, but the team was able to travel through some of the most disturbed areas to present their star shows.

The mobile planetarium used a simple spherical pinhole projector that could be slowly rotated to show the stars of all the seasons. A small cassette tape player provided a suitable musical background.

The dome was constructed like a giant five metre umbrella, it was suspended from what looked like a set of tall goalposts. Spurred by the success of the borrowed device the Society later bought a new system from Goto and successfully toured with it.

Later when he was Director of Armagh Planetarium Terence bought the first of a new type of portable planetarium, "Starlab" made by an American company.

The projector was a plastic cylinder of the pinhole type and like the old Goto could be turned to show the night sky over the year. Later Armagh used a tiny video projector to enhance the shows.

The biggest innovation of "StarLab" was the construction of the dome. It was a giant silver hemisphere that resembled a big Igloo complete with an entrance tunnel. The planetarium structure was kept in shape by air pressure provided by a continuously operated high pressure fan.

The Planetarium's Starlabs were used for outreach across Northern Ireland particularly for smaller distant schools, or community groups. Often, they were used to advertise the Planetarium itself at events and exhibitions like the "Ideal Home Show".

On a couple of occasions, the Planetarium's Starlab was displayed in the Republic of Ireland, most notably at Birr Castle, home of Lord Rosse's historic giant telescope (Figs. 13 and 14).

Fig. 13 Terence Murtagh with the newly installed 'StarLab Planetarium System' (Armagh Observatory and Planetarium)

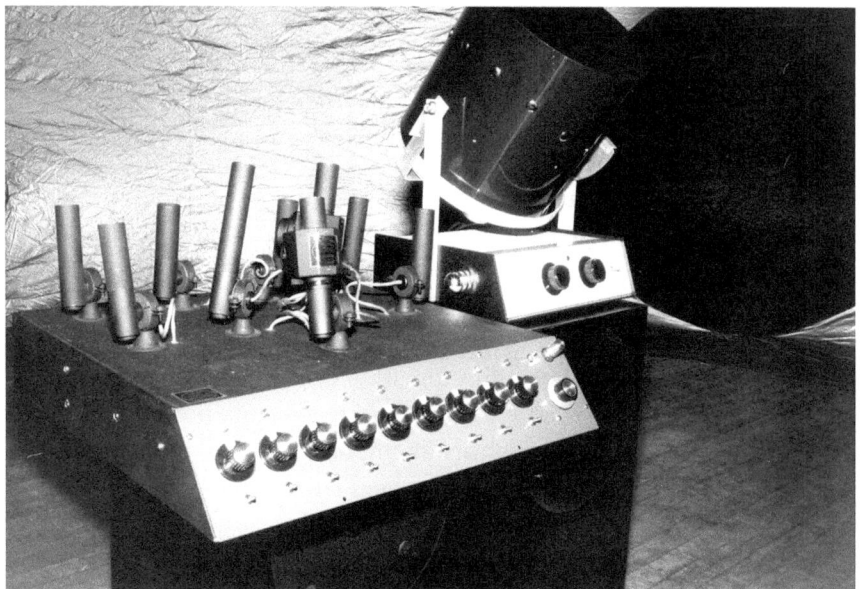

Fig. 14 Controls and special effect projector system for the portable dome (Armagh Observatory and Planetarium)

20 Planetarium Projector Museum

Owen Phairis

As we celebrate the centennial of the birth of the modern day optical/mechanical planetarium projector it seems worthwhile to also talk about the "Planetarium Projector Museum." Zeiss led the way in 1923 with the first operational optical planetarium which has followed to this day with a long line of impressive and beautiful projectors. In the late 1940s a scientific educator by the name of Armand Spitz wanted to make available an inexpensive planctarium projector that both schools and colleges could afford and so began another long line of projectors. As interest in rocketry and our space program grew in the 1960s, more companies began to produce planetarium projectors like Goto Optical, Minolta, and others. The 60's were what I refer to as the 'Golden Age' of the planetarium projector with instruments to fit every dome size and budget, but there were clouds building on the horizon.

Technology advanced as it always does, and the projectors were beginning to age. Replacement parts such as gears and lamps were becoming harder to obtain and interest began to wane. Advancements in computers and in video projection started a revolution in the planetarium. No longer would the audience be treated to watch the beautiful and fascinating scientific instrument go through its motions and recreate our wonderful night sky. The magnificent projectors standing in the middle of the dome were starting to be replaced with a box and a fisheye lens. The flexibility of the video projector that can not only produce stars, but could now take one on voyages under the ocean was replacing the large optical projector. The possibilities of this marriage between the computer and video projector seems endless. Many of the manufactures of the mechanical projectors slowed and even halted production and it appears that the future of planetariums will be mostly as video projection.

Many of the older obsolete projectors were being bulldozed and ended up in landfills or on auction blocks. I saw a need to save and preserve as many of these projectors as possible, which are an important part of our historical and scientific educational heritage. To date, the 'Planetarium Projector Museum' has collected 21 commercial planetarium projectors, the largest collection in the world, where people may come to see and learn about the evolution of these mechanical marvels that will never be built again. My goal was to collect one of each model from as many different manufactures as possible in a one-of-a-kind Museum so our planetarium history artifacts would not be lost to the ravages of time and progress. Please visit our collection in person or online at: www.PlanetariumMuseum.org (Fig. 15).

Fig. 15 Some of the Instruments held at the Planetarium Projector Museum (Owen Phairis)

21   The Nairobi Planetarium

Daniel-Chu Owen and Suzi Murabana

Science truly is for everyone. After all, it is essentially humans trying to understand our place in this vast Universe we call home. Perhaps now more than ever we need to come together as a species to fix the problems we have created due to industrialisation and globalisation. Problems which science shows us we have, but also shows us what—along with engineers—we need to do to solve them. It is now vital that everyone on Earth understands how unique, wonderful, yet fragile our planet is. Perhaps we all need the 'overview' effect—coined by Frank White, and defined as:

> … a cognitive shift reported by some astronauts while viewing the Earth from space. The most prominent common aspects … are appreciation and perception of beauty, unexpected and even overwhelming emotion, and an increased sense of connection to other people and the Earth as a whole. The effect can cause changes in the observer's self concept and value system, and can be transformative. (Wikipedia)

As of 2023, only just over 600 Earthlings have ever left our planet, and although that number will grow in the coming years it will be a long time (if ever?) before most humans routinely visit space. So, perhaps we can use planetaria to approximate this badly needed, transformative effect.

The Nairobi Planetarium began life as a clump of overgrown bamboo and an idea. Having been visiting schools all over Kenya with a large telescope and an inflatable planetarium, the Travelling Telescope wanted to build East Africa's first fixed dome. Mobiles are great, but we wanted to have something more permanent, and a fixed place where a large population of people from Kenya's capital can come by and immerse themselves in everything cosmic.

After a steep learning curve we decided to make a geodesic dome using only bamboo and rope. It took some experiments, and figuring out the best way of turning what is essentially a grass into a 10 metre diameter, 6 metre high dome. Once the main frame was complete, we hired some local "fundis" (translated as makers in Kiswahili) to stitch a waterproof cover, and then a suspended screen similar to a hemispheric sail, with each connection point attached to the fabric and pulled tight to the frame via a rudimentary pulley from ground level.

We use a powerful central projection system with a fish-eye lens, and eventually got hold of a special gaming computer with a boosted graphics processor, perfect for producing a hi-resolution, immersive environment. Most planetaria in the world are state funded, and so initial costs are borne by governments. Ours has been entirely self funded, and as such we try to stretch what we can to produce a good experience for our audience. For example, our sound system is a used, repaired high-end hifi which produces fantastic cinema-quality sound. Also, our shows comprise a mixture of selected freely available planetarium films and a variety of open source software.

It has been quite a journey, and we're still travelling!

22   Domes as Homes

Matthew McMahon

As we have seen already (The Geodesic Dome, Katie Boyce-Jacino), the planetarium brought with it advances in architecture that were crucial to the light and efficient building of domes. These domes were later picked up by an American inventor and architect, Buckminster Fuller. His philosophy of design, and a great advocate of sustainability, influenced a great deal of architects and artists in the 1950's and 1960's.[12] One such group was 'Drop City', an artistic commune on the outskirts of Trinidad, Colorado.

Drop City was founded in 1960, on a small plot of land, by a group of friends and collaborators from the University of Kansas and the University of Colorado. They were Gene Bernofsky, Clark Richert, Richard Kallweit and Joann Bernofsky. They envisioned a continuation of their earlier artistic concept, Drop Art. To create housing and work spaces they turned to the Geodesic Dome and built two, before later building two more structures based on the work of Steve Baer.[13]

Their work, along with the lectures and books by Buckminster Fuller, inspired other builders and soon geodesic domes structures were appearing all over the United States. Coinciding with the Space Race and the wave of new planetaria funded by the National Defense Education Act of 1958 and the Higher Education Facilities Act of 1963, this wave of dome building saw a rediscovery of 'The Wonder of Jena' by those building domes as homes. When Lloyd Kahn, who had written two influential books on the topic of Domes,

---

[12] Kahn and Easton (2000), p 111.
[13] Matthews (2010), p 70.

was preparing a third installment, he was assisted by Charles Hager, the Director of the Planetarium Institute at San Francisco State University.[14]

These domes prompted new companies, such as Zomeworks in Albuquerque, to be founded, which in turn developed new lightweight domes which influenced the rise and development of the portable planetarium dome, today a staple of the planetarium industry. As for the use of domes as homes, many projects continue to examine their use worldwide while early adopters such as Kahn eventually moved away from the concept and saw the dome as unsuitable for a domestic space.

23   Unisphere

Tomáš Gráf and Ondřej Smékal

The installation of a digital planetarium called Unisphere was completed at the Institute of Physics of the Silesian University in Opava in 2019. This academic planetarium is mainly used for educational purposes in programs focussing on astrophysics and of multimedia technology.

The Unisphere has an E&S (Cosm) projection system and is located in a renovated attic space in the building of the Institute of Physics of Silesian University in Opava. Its 50 seats are located under a tilted and suspended projection dome (Spitz, NanoSeam) with a diameter of 8 metres. The parameters of the projection are exceptional, at the time of installation it was the most technically advanced projection in the world, with one pixel on a spherical surface having a diameter of only 2.4 millimetres. The projection system also allows stereoscopic projection.

The Unisphere is used as an advanced teaching and creative tool, which belongs to the so-called immersive media. Its use is intended especially for students in all physics degree programmes and also for programmes aimed at communicating science to the public. Shows for secondary schools and the public are shown here several times a week, depending on interest.

One of the planned uses of the Unisphere is the actual production of full-dome shows, in which university students will also be involved. The Unisphere Studio, which is equipped with a small projection dome and another installation of the Digistar 6 projection system, is also used for this purpose.

The Studio Unisphere has already produced shows that try to explain very advanced astrophysical topics to the general public, such as binary black hole

---

[14] Kahn and Easton (2000), p 111.

systems, optical effects in extreme gravitational fields, accretion structures near black holes and the cosmic microwave background. This is a creative activity that cannot do without the close collaboration of university theoretical physicists, science popularizers and filmmakers. And this is the kind of collaboration that the team at Studio Unisphere strives for.

In the future, our own work will focus on the widest possible range of topics that promote science and critical thinking. It is also envisaged to cooperate with similar studios focused on the production of fulldome shows in the Czech Republic and abroad.

24   The Gulbenkian Planetarium in Lisbon

Pedro M. P. Raposo

In the section dedicated to the neighborhood of Belém in Lisbon, Portugal, a travel guide describes the Calouste Gulbenkian Planetarium as follows: "Financed by the Gulbenkian Foundation and built in 1965, this modern building sits incongruously beside the Jerónimos Monastery." For the occasional passerby and the many tourists who make this area one of the busiest in the Portuguese capital, the sight of a concrete cubic building topped by a dome just next to a sixteenth-century monastery might indeed look incongruent, at least at first sight. Some historical background will show that even if not visually harmonious, the placement of the planetarium is not incongruent at all.

The Planetarium, the first institution of its kind in Portugal, was established under the guidance of Naval officer and amateur astronomer Conceição Silva in the context of the relocation of the Maritime Museum of Lisbon (Museu de Marinha) to the Jerónimos Monastery. Officially named Mosteiro de Santa Maria de Belém, the Monastery was built in the heyday of Portuguese maritime expansion, following the orders of King Manuel I (1459–1521, ruled 1495–1521). The monastery is the utmost example of the so-called "manuelino style," marked by detailed embellishments that invoke seafaring and nautical deeds.[15]

The relocation of the Museu de Marinha (formerly at the Navy Arsenal) to the monastery took place during the dictatorial regime known as Estado Novo, which, resisting the wave of decolonization that followed World War II, sought to keep its domain over the colonial territories of the Portuguese

---

[15] Canas (2006), p 97–112.

Fig. 16 Façade and main entrance of the Calouste Gulbenkian Planetarium, and a view of the Planetarium and the main entrance of the Maritime Museum in the west wing of the Jerónimos Monastery (Pedro M.P. Raposo)

Empire in Africa and Asia. In order to justify this policy, the regime boasted a mythical narrative of Portugal's glorious imperial destiny grounded on its erstwhile prominence in navigation matters and nautical science.

Given the historical association between astronomy and navigation, and the symbolic importance of Belém and particularly the Monastery of Jeronimos as monuments to Portugal's history as a seafaring nation and overlord of an overseas empire, the placement the Planetarium in this area was certain to secure a place for the new institution in the life and the propaganda of the Estado Novo.

In 1974, the Estado Novo was overthrown and replaced with a democratic regime. The Portuguese overseas empire would soon be dissolved giving way to the independence of formerly colonized territories. The Gulbenkian Planetarium continued to promote public education in astronomy while becoming a meeting point for amateur astronomers and telescope makers. It remains a major scientific attraction in Lisbon—and it possibly continues to surprise tourists and occasional passersby not expecting to find a domed modern building appended to a sixteenth-century monastery! (Fig. 16).

References

Adler Planetarium, Astronomy Museum (2005) Astronomy's inspirations: a guide to art at the Adler. Adler Planetarium & Astronomy Museum, Chicago, IL

Adler Planetarium & Astronomy Museum, Schechner S (2019) Time of our lives: sundials of the Adler Planetarium. Adler Planetarium, Chicago, IL

Canas A (2006) Planetário Calouste Gulbenkian. in A Cultura na Marinha. Academia de Marinha, Lisboa

Dreyer I (n.d.) Laserium's origins. http://www.laserium.org/laserium_origins/index.html

Firebrace W (2017) Star theatre: the story of the planetarium, illustrated edition. Reaktion Books, London

Fox P (1935) Adler planetarium and astronomical museum. An account of the optical planetarium and a brief guide to the museum. The Lakeside Press, R. R. Donnelley & Sons Co., Chicago, IL

Kahn L, Easton B (2000) Shelter, 2nd edn. Shelter Publications Inc., Bolinas, CA

King HC, Millburn JR (1978) Geared to the stars: the evolution of planetariums, orreries, and astronomical clocks. University of Toronto Press, Toronto

Leonard J (25 June 1976) …Starring a concentration of light. New York Times

Leslie SW, Margolis EA (2017) Griffith observatory: Hollywood's celestial theater. Early Pop Vis Cult 15(2):227–246

Los Angeles Times (16 Nov 2019) Unusual Concert

Matthews M (2010) Droppers: America's first hippie commune, Drop City. University of Oklahoma Press, Norman, p 70

Shepard R (16 Jan 1975) A laser sheds new light on music. New York Times

Sullivan D (25 Sept 1974) Beaming in on a laser light show. The New York Times

Cultures of the Planetarium

Anna Gammon-Ross, Matthew McMahon,
Pedro M. P. Raposo, Terence Murtagh, Bing Quock,
Mike Smail, Stefano Giovanardi, Jane Kanter,
Charlotte Bigg, Susanne M. Hoffmann, Paul Cornish,
Helen Ahner, Katie Boyce-Jacino, Michael G. Neece,
David DeVorkin, and Andreas Schmidt

1   Introduction

Planetaria are host to a wide range of cultural activities and experiments. This chapter demonstrates some of the performances that have been put on in planetaria for a variety of purposes. Some have been for a didactic purpose, to better explain planetary motion, or the way in which other cultures understand the sky. Others have been to experiment with new forms of artistic expression, taking advantage of the shapes, colour and motions that are able to be produced by the planetarium projector, a domed room and human creativity.

A. Gammon-Ross
Royal Observatory Greenwich, London, UK

M. McMahon • T. Murtagh
Armagh Observatory and Planetarium, Armagh, Northern Ireland, UK

P. M. P. Raposo
Academy of Natural Sciences of Drexel University, Philadelphia, PA, USA

B. Quock
California Academy of Sciences, San Francisco, CA, USA

M. Smail • J. Kanter • K. Boyce-Jacino (✉)
Adler Planetarium, Chicago, IL, USA
e-mail: kboycejacino@adlerplanetarium.org

© The Author(s), under exclusive license to Springer Nature Switzerland AG 2024   **105**
M. McMahon et al. (eds.), *100 Years of Planetaria*, Springer Praxis Books,
https://doi.org/10.1007/978-3-031-75496-8_3

Other elements in this chapter are focused on the culture of the planetarium. The sector has developed in the last century, both in tandem and independently. The sector has been supported by organizations and publications which have been used to share knowledge, technology and experiences. Planetaria have also played host to other groups and experiments, allowing amateur telescope makers and scientists to use their facilities. Their stories are an important part of the rich cultural life of a planetarium and the sector as whole.

2   Astronomy and Islam

Anna Gammon-Ross

In November 2017, the Royal Observatory Greenwich began our Astronomy and Islam planetarium shows. This programme began in collaboration with the New Crescent Society, a Muslim community-led astronomy organization based in the UK, a group we continue to run our shows with today.

The shows cover various aspects linking astronomy and Islam. One of the main topics is how to spot the next new crescent Moon. This is important in

S. Giovanardi
Planetarium and Astronomical Museum of Rome, Rome, Italy

C. Bigg
CNRS/Centre Alexandre Koyré, Paris, France

S. M. Hoffmann
Friedrich Schiller University, Jena, Germany

P. Cornish
We The Curious, Bristol, UK

H. Ahner
Berlin-Brandenburg Academy of Sciences and Humanities, Berlin, Germany

M. G. Neece
Chapel Hill, NC, USA

D. DeVorkin
Smithsonian's National Air and Space Museum, Washington, DC, USA

A. Schmidt
Carl Zeiss Jena GmbH, Jena, Germany

Islam as it defines the calendar used to determine when specific celebrations will occur. Whenever the crescent Moon is sighted for the first time after a new Moon, then the next month in the Islamic calendar can begin. The shows also look at how Arabic astronomers have helped to shape our modern understanding of the subject, in particular the work of a man called Abd al-Rahman Al-Sufi. Al-Sufi published a work called 'Kitab al-Kawatib al-Thabit al-Musawwar' or 'Book of Fixed Stars' which cataloged the position brightness, colour, and various other aspects of the stars, expanding the work of Greek astronomer Ptolemy. It also included illustrations of the constellations, something that has helped us to not lose what the ancient Greeks imagined they saw, as well as highlighting some Arabic tales.

Over time the shows have developed and expanded to include more variations, such as a Family Special more suitable for younger children, an Astronavigation show created with help from one of our museum curators, and even a version for secondary schools.

The shows are delivered by a Royal Observatory astronomer and is followed by a short presentation from a member of the New Crescent Society which goes into more detail about the cultural aspects and practical tips for New Crescent spotting.

The shows have been very popular with almost every one being sold out! This has led on to further collaboration activities at the observatory. We have run Moon-spotting several livestreams for Ramadan and Eid, with the 2019 Ramadan stream winning an award for Best Cultural Institution at the Shorty Social Media Awards. In April 2019, we ran a special version of the Astronomy and Islam show as part of the National Hilal conference which we hosted at the museum for Imams and mosque committees.

Our planetarium closed in 2020–21 for the covid pandemic but we managed to move our Astronomy and Islam shows online, reaching audiences in their homes via Zoom. These shows continued to sell out despite losing the draw of being inside the physical planetarium building.

Imad Ahmed our lead contact with the New Crescent Society says "The partnership between the New Crescent Society, the Muslim community and the Royal Observatory Greenwich has been extremely popular. Astronomy is not just a source of wonder for the Muslim community, but it's part of our heritage, and daily ritual practice. The interest in the planetarium shows, five years on, is a testament to the ROG's success in creating in culturally relevant programming."

3   The First Planetarian

Matthew McMahon

In June 1972 the first issue of *The Planetarian* was printed and distributed to the members of the relatively young *International Society of Planetarium Educators* (ISPE) (Fig. 1). The journal was not particularly long, coming in at thirty one pages, but it contained some big names, with a central article by Isaac Asimov, and a transcript of the address by Armand Spitz at the 1970 Conference of American Planetarium Educators, his last public message to the planetarium community before his death.[1]

The origin of *The Planetarian* lay in the formation of the ISPE in 1970. While the constitution of the ISPE was being hammered out, a Publications Committee was being brought to life. The process of establishing a new journal was not easy, and there was much debate over the format of the publication, and the roles within the committee itself.[2] Eventually Frank C. Jettner emerged as the executive editor, and it was in the first 'thoughts from FCJ' that the origin of the publications name was first discussed. He credited the name *The Planetarian* to Norman Sperling, who had coined the term to describe anyone who worked in the planetarium sector.

The Planetarian was conceived to encourage the professionalization of the field and encourage collaboration between institutions. Crucially it acted as a platform for technological tips and techniques which allowed planetariums to share ideas that could be integrated into new dome shows, and to increase accessibility in the dome. Issue 2 included a segment on how Strasenburgh Planetarium was making dome shows for the deaf community, and outlined some procedures the staff felt had worked well.[3]

A news section was also included, to keep members updated on what the regional associations within the United States and Canada were organizing, and to prevent duplication of effort. By the first issue the Great Lakes Planetarium Association, the Planetarium Association of Canada, the Southeastern Planetarium Association, the Middle Atlantic Planetarium Society, the Pacific Planetarium Association and the Southwestern Association of Planetariums were all part of the 'newsbeat' segment. One of the final important features was a series written by Jettner and Con Del Chamberlain

---

[1] Spitz (1972), p 7.
[2] Marché (2005), p 165.
[3] DeGaff and Hamill (1972), p 54–55.

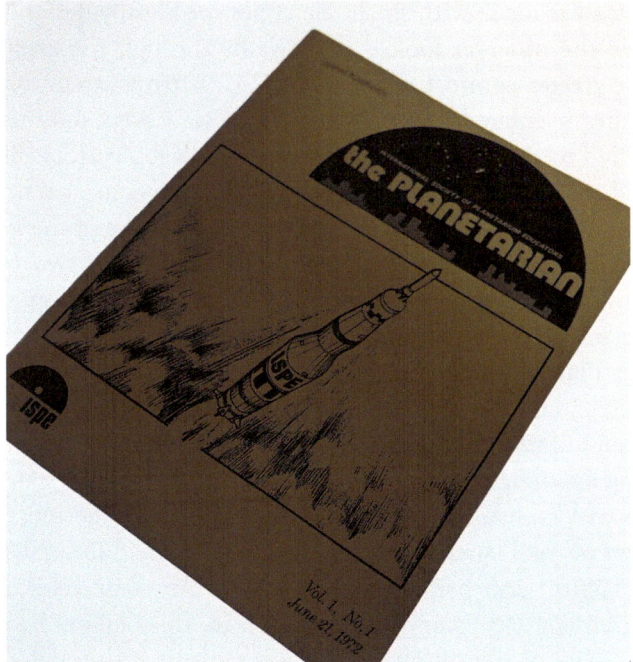

Fig. 1 The well read copy of Vol. 1, No. 1 of 'the Planetarian' from the Armagh Planetarium Library. (Armagh Observatory and Planetarium)

(Abrams Planetarium, Michigan State University) which formed the basis for an introductory handbook to planetarium operations.

4   Project Moonwatch at the Adler Planetarium

Pedro M. P. Raposo

In 1956 the Smithsonian Astrophysical Observatory launched Operation Moonwatch with the aim of tracking artificial satellites. The program was led by the astronomer Fred L. Whipple (1906–2004). It gained momentum after the launch of Sputnik I by the Soviet Union in October 1957, becoming emblematic of the Cold War and the Space Race. Moonwatch eventually counted on more than 200 teams spread across the United States and other countries.

In a typical Moonwatch session, a group of observers sat in a row along the north-south line with each individual covering a specific section of the sky.

Observations were made with small telescopes specifically designed to rest on a table while the observer looked downwards through the eyepiece, which provided for greater comfort while observing. When a satellite was spotted, the exact times it entered and left the field of view were duly recorded and then submitted to the Smithsonian Astrophysical Observatory. Besides helping track and compute the orbits of satellites, the data compiled from the various participating teams was used to refine the knowledge of the Earth's shape and atmosphere, which had a particular relevance as the two major world powers, the United States and the Soviet Union, were developing their ballistic missile arsenals.[4]

The Adler Planetarium soon made its observing deck available to young Project Moonwatch observers from the Chicago Astronomical Society (CAS). CAS had been founded in 1862 by a group of Chicago civic leaders interested in promoting astronomy as a way to boost the city's cultural status and prestige. Efforts to establish a solid partnership between Adler and CAS during the early days of the Planetarium back in the 1930s had limited success. But by the late 1950s, CAS had absorbed the local Amateur Telescope Making (ATM) movement (see story about ATM in this volume), and Project Moonwatch had a significant resonance with it. Although a standard Moonwatch telescope was sold by the Edmond Scientific Corporation, many volunteers made their own instruments following the project's guidelines. Adler was already a hub for ATM in the Chicagoland area, thus opening its doors to Project Moonwatch reinforced its ties with the local amateur community.[5]

It also showed that the Planetarium was embracing the Space Age, in contrast with an alleged conservative attitude towards space flight in the years that preceded the launch of Sputnik I. In his recollections of the time he spent at the Adler around 1954, while still a high school student, physicist and space scientist George Carruthers noted that spaceflight and the possibility of doing astronomy from space (ideas that Carruthers had encountered in an article by Fred Whipple in *Collier's* magazine) were deemed nonsense by the Planetarium's scientists. By the late 1950s, the Adler's approach to space exploration was inevitably changing, and its keen support to local Project Moonwatch volunteers can be seen as an early effort towards refashioning the institution as a planetarium for the Space Age (Fig. 2).[6]

---

[4] Bolt (2009).
[5] McCray ().
[6] American Institute of Physics (2015).

Fig. 2 Young Project Moonwatch observers from the Chicago Astronomical Society in action at the Adler Planetarium, c. 1957–58 (Adler Planetarium Archives)

5   Making Telescopes at the Adler Planetarium

Pedro M. P. Raposo

In the late 1920s, a movement known as Amateur Telescope Making (ATM) started to take shape in the United States. Amateurs all over the country (and eventually, the world) formed groups and associations devoted to the making of telescopes. The leading figure in this movement was Albert G. Ingalls (1888–1958), who edited a column in *Scientific American* titled 'The Amateur Scientist'. Ingalls was also the author of *Amateur Telescope Making*, a work that went through several editions and which was crucial in promoting amateur astronomy in the USA and around the globe.[7]

The ATM movement soon gained a footing in the Chicagoland area. The Amateur Telescope Makers of Chicago (ATMC) were led by Arthur Howe Carpenter (1877–1956), a professor of metallurgy at the Armour Institute in Chicago (now the Illinois Institute of Technology). Two other leading figures

---

[7] Buttles (1932), Callum (1936), Cameron (2010), Williams (2000).

in Chicago amateur astronomy were Lois and William Buttles, a couple that established a separate amateur society named Astrolab. Astrolab attracted members from all over the United States and other countries around the idea of collaboratively constructing standardized portable telescopes that were then taken to schools, community groups, and other organizations for public observing programs.

The astronomer Philip Fox, who led the Adler Planetarium between 1930 and 1937, recognized that engaging with Chicago's vibrant amateur community would help build a civic basis of support and consolidate a regular audience for the Planetarium. By opening the Planetarium's doors to this community and inviting amateurs to use the Planetarium's premises and equipment (which included an optical shop and a Zeiss polishing machine), Fox turned telescope making into one of the hallmarks of the Adler's activity as an educational institution for the ensuing decades.

By the early 1950s, the optical shop occupied an important role in the Planetarium's offerings, hosting telescope-making sessions where amateurs and enthusiasts could grind their own telescope mirrors and assemble their instruments under the guidance of Planetarium staff and fellow amateurs. Around 1960, the optical shop became a "living" exhibition titled "Amateur Telescope Makers Optical Shop." Visitors walking by could watch staff and participants engaging in telescopic making activities, or at least get a glimpse of the shop's set up and tools when not in use. The Planetarium further promoted its liaison with the ATM movement by hosting a Telescope Makers' Fair.

The optical shop was dismantled in 1999, but the Planetarium has continued to promote engagement with telescopes and astronomical observation through public observing programs and events, both in and around the Planetarium's premises and by taking portable telescopes to local libraries and community centers around the Chicagoland area (Fig. 3).

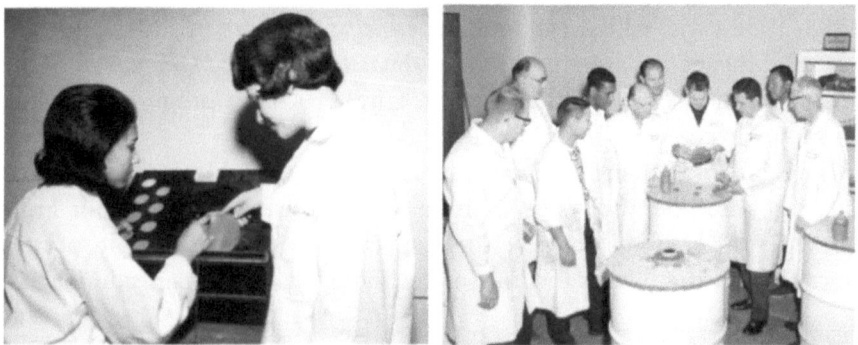

Fig. 3 Staff and amateur telescope makers at the Adler Planetarium's optical shop, c. 1960 (Adler Planetarium archives)

6   Astronomical Research as an Inspiration for Modern Art

Andreas Schmidt

The planetarium in its modern appearance has recently become more than just a training facility for astronomers, navigators, and astronauts. Thanks to modern digital technology, they have long since become places of cultural encounter and immersive experiences. Impressively, you can catch sight of entire galaxies, star clusters and nebulae in multiple colors and highest resolution. To put it briefly, all the wonders our universe has to offer. An experience that has been shaped not least for decades by space telescopes such as Hubble and has thus become the expectation of the modern public.

However, it is often not the scientific background that is in mind of the visitors, but rather an enthusiasm for the overwhelming aesthetics. Amazement is an important part of the experience. Undoubtedly, this aspect also helps to consolidate knowledge, because it is much easier to learn when knowledge is combined with emotions.[8]

Inside the planetarium the universe presents itself as an ingenious creator of magnificent shapes in an often-tremendous interaction of colors and fascinating dynamics. Of course, all this is enabled by the tireless ambition of the many hardworking and creative digital artists. By them the enormous dimensions of cosmic structures are visibly scaled down, their components are broken down into colors schemes, cosmic time scales are reduced to seconds. It is solely due to this deliberate adjustment, as well as the use of now-common dynamic simulations and three-dimensional models, so that the impression of plasticity and structure becomes tangible. Nevertheless, the experience remains overwhelming, and for some the sight is beyond measure. For the rest, at least, it takes their breath, which often leads to a dreamy and drawn-out "wow". Therefore, the pictorial representation of our whole universe has long had the potential to inspire the fine arts.

Anyone who feels inspired to bring the impression of wonder and fascination from the planetarium to his/her/their living room by means of artistic painting, does not need to be particularly trained or talented. As often, it is good to know how it works. The recommended technique is called "pouring", in which acrylic paint is diluted with a special medium and then poured onto the image carrier. The gravity does the rest. The result obtained, depending on their own willingness to experiment and the color combinations used, can certainly be understood as abstract art. One recognizes dynamic currents and

---

[8] Schmidt (2020).

Fig. 4 Andreas Schmidt, Neptune (2020, acryl on canvas, 40 × 50cm, private collection)

vortices, banished onto a canvas. Perhaps in shades of orange. To your astonishment it will appear almost like a photo of Jupiter clouds of the Juno space probe. Of course, you have every choice of any color. How about blue? Blue like the planet Neptune, which is according to modern research even more dynamic than Jupiter. Should a space probe pass again someday the Neptune to reveal new fantastic details, then this should be quite similar to what has been decorating your own living room for years (Fig. 4).

7   The Planetarium as Museum

Pedro M. P. Raposo

The Adler Planetarium and Astronomical Museum opened to the public on May 12, 1930, to coincide with the birthday of its benefactor Max Adler. It was built in what is known nowadays as the Chicago Museum Campus, next to the Field Museum and the Shedd Aquarium. In the words of the Planetarium's first director, the astronomer Philip Fox, the three museums formed a "trinity: the heavens above, the Earth beneath, and the waters under the Earth."[9]

Together with his assistant, the astronomer Maude Bennot (see story in this volume), Fox played a leading role in shaping the planetarium's original programs and displays. He had traveled to Europe in the summer of 1929 to visit museums and planetariums. It was apparently during those travels that Fox became interested in antique scientific instruments and started to develop a vision for the new planetarium in Chicago as a full-fledged museum of astronomy, featuring the original Zeiss Mk II projector as the main attraction.

Fox successfully persuaded Adler to also sponsor the purchase of the so-called Mensing Collection from the Dutch antiques dealer Anton Mensing. It comprised c. 500 historical scientific instruments, including sundials, telescopes, astrolabes, globes, surveying instruments, and other devices. The Mensing collection was used in a series of displays around the Planetarium's original dome theater. They consisted of cabinets showing the instruments more or less organized by type and function. The cabinets were complemented with displays featuring astronomical images captured at the leading observatories of the time (such as Yerkes, Mount Wilson, Lowell, Greenwich, and Meudon) and exhibits highlighting contributions of Chicago's institutions and scientists to modern astronomy.[10]

The "Astronomical Museum", later "Astronomy Museum" part has long been dropped from the Adler's official name, but the Planetarium continues to display historical objects from the Mensing collection and others added over the decades, in innovative exhibitions telling the stories of how people of all backgrounds have engaged with the skies over the centuries, effectively bringing Fox's and Bennot's original concept of a museum of astronomy into the twenty-first century (Fig. 5).[11]

---

[9] Fox (1935), p 26.
[10] Korey (2012), p 161–7.
[11] Taub (1995), p 243–250.

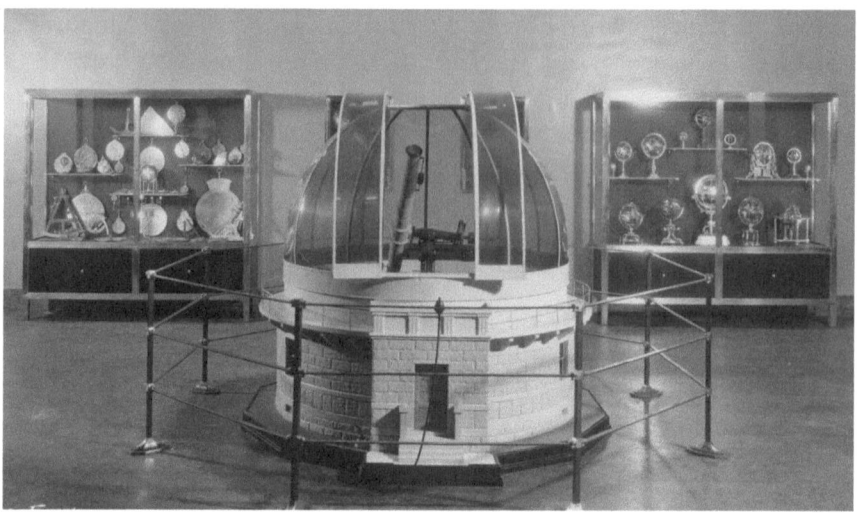

Fig. 5 Some of the original museum displays at the Adler Planetarium c. 1932, with a model of a modern observatory on the foreground and displays of historical scientific instruments in the background (Adler Planetarium archives)

8   Bottle Rocket Workshop

Terence Murtagh and Matthew McMahon

Whilst the dome has always been the focus of a planetarium, they have rarely been the only experience on offer. Exhibitions, talks and workshops have been part of the planetarium visit for decades, each aiming to cater to a different aspect of the experience. The traditional dome show, with the accompanying presenter-led talk, is now supplemented by other activities.

In Armagh Planetarium one such activity is the Bottle Rocket workshop. This workshop has been running since the mid 1990's and was developed as part of a summer package to help with the flow of guests through the exhibition area and to the dome. However, it proved so popular that over twenty-five years later it is still a major part of the planetarium visit.

Essentially a scaled-up home experiment, it aims to give children an introduction to aerodynamics, and basic rocketry. Compressed air and water as a propulsion method and bottle for the rocket serve as a canvas for creative designs, experimentation, and spectacle. Children are encouraged to ask why rockets are built in certain shapes, and to discuss the challenges in groups. Even color and design enters the scene as groups learn about why spaceships are stamped with flags and given names. The spectacle comes with the finale as the crafted rocket is launched into the air with an explosion of water and sound, exactly what children expect a rocket should do!

Fig. 6 Children launching a bottle rocket at a school in the early 2000's (Armagh Observatory and Planetarium)

Under the surface the workshop, like so many of the activities run by planetaria all over the world, serves the education of the public. Over the last century the planetarium has begun to cater to more and more different types of learning style, encouraging conversational learning, kinesthetic play focused experiences and musical events. Today's planetarian is an educational professional with an understanding of not just astronomy, but educational theory and accessibility. Workshops serve to open the dome to a wider audience, encouraging them to interact with space science in new and interesting ways (Fig. 6).

9   The Vortex Concerts

Bing Quock

A few years after San Francisco's Morrison Planetarium opened in 1952, sound artist Henry Jacobs and filmmaker Jordan Belson, intrigued by the medium's potential, decided to think outside the box (or, rather, outside the dome) and create what Belson described as a new kind of "non-intellectual, non-educational and non-referential" form of communication. In 1957, they proposed an experimental demonstration to the California Academy of Sciences, home of the Morrison Planetarium.

The circular arrangement of the planetarium's audio system was of particular interest to Jacobs, and with help from outside engineers, he devised a rotary controller with which he could direct sound to any of the theater's multiple speakers, located along the perimeter of the dome. From the program notes of the third series in 1958: "The name—Vortex—is derived from the ability to move the sound around the dome in either clockwise or counterclockwise rotation, at any speed. According to the *San Francisco Tape Music Center: 1960s Counterculture and the Avant-Garde (edited by David W. Bernstein)*, "The objective was to immerse the audience in a virtual whirlwind, a 'vortex' of sound and light."

To accompany the sound, Belson created abstract visuals that were largely circular or radial, including interference (or moiré) patterns, spirals, and mandalas. The enveloping expanse of the planetarium dome produced a frameless, open canvas against which Belson's images seemed to float hypnotically around the audience, produced by up to 30 projection units. "We were able to project images over the entire dome," he said, "so that things would come pouring down from the center, sliding along the walls."

The Vortex Concerts ultimately proved to be so popular that Jacobs and Belson were forced to perform additional shows. As reported by Jordan Marché in *Theaters of Time and Space: American Planetaria, 1930–1970*, Planetarium director George Bunton was concerned about the composition of the audience, "roughly one-half of whom he described as 'Bohemian,' or belonging to the beat generation and widely regarded as 'denying themselves nothing in the way of worldly pleasures and sensations.' According to Jacobs in a 1994 interview on KPFA, they had drawn "too many people smoking funny-smelling cigarettes outside in the line." Interestingly, later light shows would use their attractiveness to non-traditional audiences as a major selling point to museum administrations looking to boost attendance.

Unfortunately, no film record of the Vortex Concerts exists, since the projections in the shows were too faint to register. However, Belson later produced several short films that incorporated concepts that were used in the planetarium, and audio highlights were released on vinyl and have been preserved digitally. These are the only surviving traces of the original Vortex Concerts, which paved the way for all light shows that followed, from rock concerts to indoor and outdoor laser shows to fulldome computer-generated digital displays. Jacobs and Belson were truly pioneering fathers of the abstract planetarium light show experience!

10   The Planetarium Lecture Series

Mike Smail

About a decade after their revolutionary Vortex concerts (The Vortex Concerts, Bing Quock), San Francisco's Morrison Planetarium started another musically-tinged project, the Planetarium Lecture Series. This was 'an attempt to bring the thrill and intellectual stimulation of planetarium lectures into the home and classroom... through dramatic recorded adaptations of the kind of programs currently offered in planetariums everywhere.'

Morrison Planetarium lecturer Hubert J. Bernhard was the author of these presentations. Bernhard was a former San Francisco newspaper reporter and editor, as well as an author of several astronomical texts, and even a science fiction short story titled 'Welcome Voyagers' about Earth's first contact with an alien species. In over 20 years with the Morrison, Bernhard would present over 3000 public performances!

The primary series consisted of five Bernhard lectures: *Life Among the Stars*, *The Christmas Star*, *The UFOs*, *Mysteries of Mars*, and *The Other Moons*, all released between 1966 and 1970. The planetarium collaborated with Optica, an Oakland-based company who created the Stellar Ventures label to release and distribute these lectures to the public. Bernhard's narration was paired with an underlying soundtrack of bells, drums, and often very dramatic organ music that may or may not have been part of the lectures at the planetarium. At the time, each of the five lectures was available on vinyl record ($4.00), cassette tape ($6.95), and reel-to-reel tape ($9.95). Nowadays, you'll only find the vinyl records on secondary markets like Discogs and eBay. Each record is available in a black and white or single-color cardboard jacket, for the true collecting completists among you.

Each lecture is approximately 40 minutes long, and Bernhard's storytelling holds up, even by modern standards. While the longer running time afforded him plenty of opportunity to share minor scientific details, he was also very up front about all the things that we don't know, giving audiences plenty of questions to think and theorize about for themselves. On *The Other Moons* record jacket, listeners were even invited to send in a self-addressed, stamped envelope to receive a free test (with answers) for either self-review or to test a class.

Despite the focus on scientific knowledge, quite a bit of science fiction, or science speculation, permeates the lectures. While there's an entire lecture dedicated to UFOs (which Bernhard delightfully pronounced YOU-foes), and possible explanations for the Star of Bethlehem, there is also discussion in the other lectures about the likelihood of life existing on Mars or elsewhere in the universe, and the possibility of moons being hollow, alien space stations.

While the main five lectures of the series are easy to find, even today, many people are unaware that a second series was planned and partially-released. In 1973, Bernhard and Optica copyrighted a lecture titled 'Taurus the Bull—Zodiacal Constellations', which the record jacket lists as Lecture Number Z-3. This lecture opens with tales of the constellation in mythology, followed by a tour of Taurus' astronomical highlights: the Crab Nebula, Pleiades, Hyades, Theta Tauri, and more. The soundtrack is notable more modern then the first five releases, featuring a wandering guitar line throughout the previously-familiar instrumentation. What were lectures Z-1 and Z-2? Were all 12 lectures ever released? We don't know. Taurus was the only zodiac lecture with US copyright filed by Bernhard and Optica. And the copy of Taurus in the author's personal collection is the only one they are aware of. If you've seen others, the author would love to hear from you! In the 50 years since its release, several other composers and planetariums have publicly released sky show recordings and soundtracks on vinyl record and compact disc, but there has yet to be another sustained effort like the Planetarium Lecture Series.

11   An Asteroid for a Planetarium

Stefano Giovanardi

In 2005 the Planetarium of Rome became the first in the world to have an asteroid dedicated by the International Astronomical Union: (66458) *Romaplanetario*, discovered by renowned Italian astrophysicist and Planetarium staff member Gianluca Masi in August 1999, using his private "Bellatrix" observatory in Ceccano, near Frosinone, about 100 km southeast of Rome. Masi was searching for asteroids in the constellation Pisces—the Fish—, when he noticed one that was apparently moving in the opposite direction to regular asteroids. The object was identified as a new minor planet, and labeled 1999 QV1: it was initially suspected to be a Near Earth Object, however further observations showed that it is safely orbiting in the Main Asteroid Belt—390 million km from the Sun–, on a rather eccentric ellipse, with a period of 4,2 years.

In the following years the object was followed up until its orbit was precisely determined; then Masi was allowed—as the discoverer—to suggest a motivated designation for the asteroid name. Having heard of the work in progress to rebuild in Rome the Planetarium—that had been closed 23 years before—Masi decided to dedicate his new asteroid to the new Planetarium of Rome, to celebrate with a celestial dedication its important return in the Eternal City.

The designation of asteroid 1999 QV1 as (66458) *Romaplanetario* was officially approved by the IAU Commission in January 2005, with the following motivation:

> "Without a planetarium for more than 22 years, the Everlasting City now has a modern one, opened in 2004. Hosted at the Museo della Civiltà Romana it has a 14-meter dome. The discoverer contributes with the staff to introduce visitors to the wonders of the universe." With this dedication the international astronomical community acknowledged the fundamental role of a Planetarium in the diffusion of scientific culture to the public.

On the second anniversary of the re-opening of the Planetarium of Rome, in 2006, an artistic model of the asteroid *Romaplanetario* was presented to the public, featuring a scaled replica of central Rome on its surface. It accurately represents in bas-relief the streets, the monuments, the city walls and the Tiber river, based on the estimated size of the object, approximately 3 km, ranging from St Peters to St John's basilicas.

The model of asteroid *Romaplanetario* followed the displacement of the Planetarium of Rome to several locations between 2014 and 2022, due to renovation works at the Museum of Roman Civilization. Since the re-opening of the Planetarium in April 2022 it is finally on permanent display at the center of the Planetarium dome: *Romaplanetario* has become the Planetarium of Rome's "good star".

12   Arcturus Lights the Century of Progress Exposition in Chicago

Pedro M. P. Raposo and Jane Kanter

In the evening of May 27, 1933, a crowd of 30,000 gathered in what is nowadays Northerly Island in Chicago for the opening of the Century of Progress Exposition. The fair covered the city's southern lakefront, including the recently-constructed land known as Northerly Island, with dozens of buildings and more than a hundred attractions—including the Adler Planetarium, opened just three years earlier. The crowd stood in an open-air court in front of the Hall of Science, one of the main buildings in the Exposition. The sound of the festivities was also disseminated throughout the fair grounds and broadcast on the radio. The proceedings were led by the astronomer Philip Fox, the director of the nearby Adler Planetarium and Astronomical Museum.

After a classical music performance and speeches by Rufus C. Dawes (a prominent Chicago businessman and the head of the Century of Progress enterprise), Edwin B. Frost (former director of the University of Chicago's Yerkes Observatory in Williams Bay, Wisconsin), and Philip Fox, the Adler's first director, the attending crowd was presented with a spectacular stunt that connected earth and sky. An illuminated panel above a rostrum showed a map with the locations of four astronomical observatories—Yerkes, Harvard, Allegheny (Pittsburgh), and University of Illinois (Urbana). After each observatory was pronounced ready, the respective mark was encircled on the map. Then a star appeared in the center of a flaming circle. Finally, a switch was thrown by Fox, triggering a searchlight at the top of the Hall of Science. The white beam started to move slowly from one building to another. Upon being touched by the beam, each building burst into a colorful display of light.

The electric current that triggered the show had originated in space, more specifically in the star Arcturus, one of the brightest and easiest stars to identify in the spring sky from Chicago's latitude. Its light had been captured at Yerkes Observatory (with the other three observatories acting as a back-up) with a photoelectric cell attached to Yerkes' 40-inch refracting telescope. The resulting electric signal was then relayed to the Century of Progress grounds via telegraph wires. At the time, Arcturus' distance was reckoned to be 40 light years, meaning that the light used to illuminate Century of Progress had left the star while Chicago was hosting its first world fair, the Columbian Exposition of 1893.

The Arcturus stunt proved so popular that it was allegedly repeated every night for the duration of the fair. It showed that astronomers were able to harness the forces of nature and capture the energy of stars in a time when electricity was becoming a hallmark of modern life and photoelectric cells were regarded as a state-of-the-art technology. This was particularly significant since astronomy, historically the basis for applied sciences such as

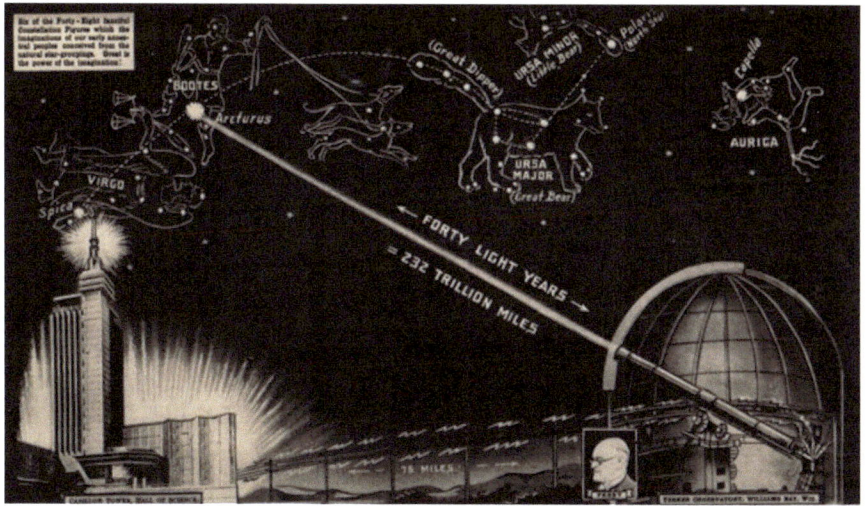

Fig. 7 The Arcturus Stunt explained in Norton Wagner, *Unveiling the Universe* (Scarbtom, PA: the Research Publishers, 1936), Adler Planetarium library

timekeeping, navigation, and surveying, seemed increasingly focused on probing the far recesses of the cosmos and removed from mundane affairs. Of course, it was also a good promotion for the Adler Planetarium, the first institution of its kind in the Western Hemisphere and a major attraction of Century of Progress.

"It is difficult to imagine," Fox later stated, "any experiment which could more vividly or dramatically illustrate the new knowledge [of the field of astrophysics]." Yet he was quick to note the deep foundations of this field—foundations which could be readily seen in Adler's collections of astronomical instruments. We might be able to power our technological marvels with stars, but no matter how new the technology, the starlight itself will always be old—making real and visual connections to both the past and the sky.

Similar performances involving cosmic rays and radio signals from a supernova were later used to open the world fairs in New York (1939) and Seattle (1962) respectively, but the Arcturus stunt occupies a special place in the history of planetariums given its close connection with the early history of the first modern planetarium in the Americas (Fig. 7).[12]

---

[12] Osterbrock (1997), Havlik (2006), Ganz (2008).

13   Spectacular Astronomy

Charlotte Bigg

Since the early twentieth century, planetarium shows have become the astronomical spectacle par excellence. Yet the history of putting astronomy before audiences using spectacular means goes back further in time, as it has continued to evolve into our days, shaped by scientific, technological and cultural innovations.

In 2014–2018, I was involved in bringing together an international, interdisciplinary group of researchers from the human, social, and exact sciences as well as artists, visual technicians and planetarium and museum professionals to investigate the history, present state and future of popular astronomical spectacles. The first leg of the project was historical and explored astronomical spectacles as productions and experiences at the intersection of theatrical and performance cultures and of the history of science and popular science. The second leg of the project was experimental, involving artists and planetarium professionals building on this historical research and using new immersive technologies to imagine alternative, possible planetarium shows. Thereby we aimed at stimulating a reflection on the science, techniques and technologies both thematized by and deployed in astronomical shows. Using this dual approach, we studied for instance human planetariums, where individuals are made to re-enact the motion of the solar system's planets. Human planetariums are recorded in history as a courtly past time and they practised today in pedagogical settings. Could a virtual human planetarium help us understand past astronomical performances, think about current pedagogical approaches to teaching astronomy, and explore the experiential possibilities of virtual reality? The collaboration of planetarium and museum professionals and academic researchers brought many new insights that benefitted both research and practice.

The main outcomes of the project were:

- Journal special issue including contributions by project members and guests: Bigg C, Vanhoutte K (eds) (2017) Spectacular Astronomy, special issue of Early Popular Visual Culture 15/2:115–272
- Artistic performance (immersive planetarium): Eric Joris and Crew, Celestial Bodies:

- https://crew.brussels/en/productions/celestial-bodies
- Artistic performance (theatrical play written for planetariums): Pieter de Buysser, The Tip of the Tongue: https://www.pieterdebuysser.com/en/work/yearly-overview/54-werk/2017/252-the-tip-of-the-tongue

The project was funded by the University of Lille III and the French Ministry of Culture and Communication (Figs. 8 and 9).

Fig. 8 Mr. Perini's Planetary in London. Anon (1880) Un nouveau planétaire. La Nature 8/2:17–18, on 17 (Charlotte Bigg)

Fig. 9 Charlotte Bigg and her avatar, acting as a guide for immersants in Eric Joris and Crew's virtual human planetarium (Charlotte Bigg)

14   Wanna Learn Constellations?

Susanne M. Hoffmann

Look, there is a pattern of seven rather bright stars: what do you prefer to see there? Some see a plough, others a casserole, others a wagon or an elk. These seven stars form only a part of the official constellations of Ursa Major, the Great She-Bear.

See, the constellations are not natural phenomena. The stars are not physically connected, they only occasionally stand next to each other in the sky for the observer on planet Earth. All cultures make their own constellations: for people in the far north who live with the caribou (e.g. the Inuit in the northernmost areas of America and the coast of Greenland), a deer might appear more obvious in the specific constellation of dots than for a Mediterranean person (e.g. the ancient Greeks) who might have called the figure a bear—not because it is easily recognisable but because it points the direction towards the "land of the bears", the north. Recognisable are, however, other cultures' imaginations such as the Babylonian "Wagon" or the Chinese "Northern Dipper". The reason and the cultural context for all these people were certainly different.

When constellations are not "real": why should we care about phantasies of our ancestors?

If we abstain from telling the stories that are public known as "Greek" but go back to the real origin of the constellation, we will always find some reasons in nature. The stories that are publicly narrated as "Greek" mythology, are—in most cases—Roman variants of earlier originals. Additionally, it has been proven by historians (Doumanis 2010) that "the Greek" is in fact an artificially made-up culture in order to unify the many cultures in Alexander the Great's huge empire. The "Greek" constellations that had been fixed in Hellenistic times consist of Babylonian, Persian, Egyptian, Israeli, Syrian, and Hittite influences, mixed with some cultural context from earlier Greek city states like Athens or Sparta. Going back to their real origins, we will find much more than some more fairy tales from other cultures: we will find patterns to navigate in time and space. The constellations tell navigators on seas and in deserts their way but they also tell farmers and sailors when to begin specific seasonal activities. For instance, the term "Pleiades" for a star cluster originates from the ancient Greek term "pleion" (to sail) because the star cluster's heliacal rise is at the beginning of the sailing seasons and its opposition (ancient term: achronycal rising) designates the end of sailing season. The sailing season depends on the predominant wind direction whose changes depend on the seasonal shift of the trade winds which, in turn, depend on the sun's position (above the northern or southern hemisphere). Therefore, teaching constellations and their historical background, frequently leads to questions that require a profound understanding of the Earth, its geology, meteorology and climate.

Similar effects can be discovered for any culture on Earth: the indigenous peoples in Australia and New Zealand will also find themselves influences by the (southern) trade winds. The peoples in the far north—no matter if in Asia, America or Europe—will always be effected by the midnight sun and seasonal periods of long darkness (or twilight). The more we understand the cultural naming of the patterns in the sky by our ancestors, the more we will discover the natural environment of the climate zone in which we live.

To me, this appears quite appealing and a very good modern way of teaching constellations: let's not focus on the funny stories of forgotten old cultures (although remembering them might trigger our memorics of the patterns) but on the practical purpose that constellations were made for!

From the early beginnings on, planetariums have been used to show the sky—but not only the local sky but also the sky above other places on our wonderful planet. The great chance of imagining to travel in a comfortable seat in an air-conditioned room without all the stress of heavy luggage, missed trains, ships or planes, or adaption to another time zone or the different temperature is definitely one of the advantages that we can use modern fulldome-equipped planetarium domes for. We can show wonderful documentaries of places, animals and plants of other continents, and we can correctly display the sky above the indigenous humans living at these places and using the stars, forming constellations for their orientation in time and space.

Interestingly, the original concepts of making constellations have been the same all over the world: As for the Australians, the dark constellations indicate the time for emu nest predators, the Babylonians had a group of constellations made of stars that indicate the season for ploughing the ackers. As for the Khoikhoi and San in South Africa, also for the Navajo in North America and for the Babylonians, all stars are part of a creation myth. It really doesn't matter where exactly we are based on Earth: all humans had basically similar ideas but with different performances depending on their local cultures that fits the local seasonal patterns. Thus, learning constellations can also further the understanding that we are all humans (no matter of skin-deep differences).

Let's go to pathfinder- or explorer-mode! Let's do it like our ancestors and use the constellations to identify the seasons and for forecast the seasonal climate (if not even weekly weather if connected to the global wind cell system).

15   Scale Models and Extraterrestrials

Pedro M. P. Raposo

Among the exhibits most commonly found in and around planetariums and science museums all over the world are scale models showing the sizes of the planets of the solar system in relation to that of the sun. A well-known example is that of the Rose Center for Earth and Space of the American Museum of Natural History in New York City. The Adler Planetarium in Chicago also sports a similar exhibit inside its Sky Pavilion, shown in Fig. 1. The top of the dome of one of its sky theaters protruding from a lower level of the building provides a reference for the size of the sun, while the gas giants, too large to be displayed on the floor, hang from the enclosing glass and metal structure above the exhibition space.[13]

Other displays, usually installed outdoors, show the orbits of the planets in their proportional sizes. In some cases, sizes of the solar system bodies and

---

[13] Raposo and Graney (forthcoming).

sizes of orbits are combined into a single exhibition according to the same scale factor. Because astronomical distances are astoundingly large, these models must inevitably extend over large distances. Notable examples in the United States include a 40-mile (64 kilometer) long model from the sun to Pluto installed in Aroostook County, Northern Maine in 2003, and the Peoria Riverfront Museum's Central Illinois Community Solar System, which includes comet markers and takes the Alpha Centauri system (the closest star system to the sun) to be represented by a moon crater in the Apollo 11 landing site! By superimposing proportional representations of astronomical sizes and distances to familiar areas and landscapes, these models, particularly when represented on a map, convey a dramatic impression of how large astronomical distances are—or, in other words, how empty space overall is.

These exhibits may be regarded as extravaganzas of modern science communication, but the methods they employ to represent astronomical sizes and distances have a long history. They were mainly codified in a book by the seventeenth-century Dutch astronomer and physicist Christiaan Huygens titled Kosmotheoros and published posthumously in 1698. The title of the English version that was published in the same year, Celestial Worlds Discover'd, or, Conjectures Concerning the Inhabitants, Plants and Productions of the Worlds on the Planets, leaves no doubt as to the subject matter and intent of the book. Huygens argues for the plausibility of a solar system teeming with life and of each "fixed star" having its own inhabited planetary system. In order to persuade the reader, Huygens produced several diagrams including the one shown in Fig. 2 in chapter "Objects of the Planetarium". It compares the relative sizes of the planets of the solar system known to Huygens to that of the sun. Huygens intended to show that the Earth, included in the diagram, was a relatively small and indistinct planet in terms of size, and that it would be difficult to conceive the waste of matter and space if all other worlds were not inhabited. Another diagram illustrates a similar argument but by showing the relative sizes of orbits and how empty overall the solar system is. While Huygens expanded on ideas and visual techniques put forward and used by other authors before him, his diagrams have been imitated and adapted countless times ever since, eventually jumping out of the pages of astronomy books to planetarium galleries and even whole cities, regions, and beyond!

Next time you see such a display, take a moment to appreciate how incredibly large the universe and how remarkable this small planet of ours are—and also how the ingenuity of a seventeenth-century scholar fascinated by the idea of extraterrestrial life still continues to inform the ways in which we communicate and visualize the immensity of the cosmos (Fig. 10).

Fig. 10 Scale model of the planets of the solar system at the Adler Planetarium (Adler Planetarium)

16   Researching Indigenous Star Stories

Paul Cornish

> According to a Native American story, the Summer Triangle represents a hand with a pointing finger. As Summer ends, the finger points south, telling everyone to head for warmer lands before Winter arrives.

I was told this story in 2005, and in the years that followed I enjoyed dropping it into my Planetarium presentations whenever the opportunity arose.

By 2022 I was working at the We the Curious Planetarium. We were searching for alternatives to Ancient Greek myths, and I remembered the Summer Triangle story. It seemed to fit the bill, but this time I wasn't just presenting a show, I was writing it. I wanted to ensure I wasn't talking out of my hat.

While verifying a story for a previous show I had contacted Roger J. Lewis, curator at the Nova Scotia Museum. Mr Lewis had kindly confirmed that I was telling an authentic Mi'kmaq tale. He told me that when sharing the story, I should emphasise that it's used for teaching purposes in a culture based on oral narratives and I should not trivialise it.

Mr Lewis' advice informed my approach to the Summer Triangle story. I wanted to ensure I was telling it respectfully and in the correct context. Unfortunately, I had no idea to whom it belonged. The only place it seemed to exist was on a website called starlab.com. I contacted them and was informed that the story had been uploaded years ago, and nobody currently working there had any knowledge of its source.

Starlab did however, direct me to Native Skywatchers, a program designed by Dr Annette S. Lee to record and share Indigenous star knowledge. Nativeskywatchers.com, is a fantastic resource, containing recordings of seminars given by astronomers and indigenous cultural experts. I learned a great deal from the work of Dr Lee and her colleagues, but I was unable to find anything resembling the Summer Triangle story.

Through the National Museum of the American Indian I was put in touch with Michael Conolly Miskwish of the Kumeyaay Nation. He recommended his book to me—*Maay Uuyow*.

The cosmological knowledge of the Kumeyaay was suppressed by Spanish invaders in the 1700s. Mr Miskwish has pieced this knowledge together using anthropological research and what has survived through Kumeyaay oral tradition.

His book contains many beautiful stories, but the tale of the Summer Triangle remains elusive. It may well be an authentic story that has only ever been passed down orally, but without a reliable source I decided not to include it in our show.

Despite this, I found the process of seeking out the people to whom these stories belong to be rewarding in itself. It gave me the opportunity to experience even more stories and taught me more about cultures other than my own.

And while I may still occasionally find myself talking out of my hat, I can at least ensure that I am doing justice to these wonderful stories (Fig. 11).

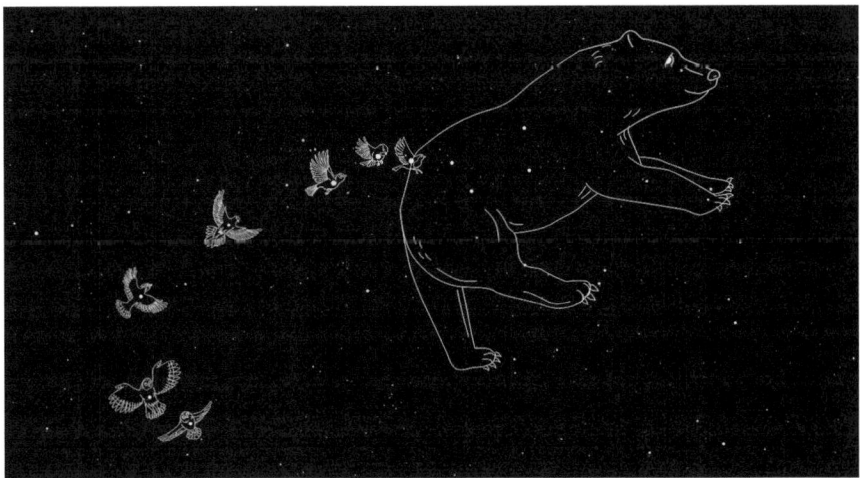

Fig. 11 Mi'kmaq story of Ursa Major—Muin the Celestial Bear (Evans and Sutherland)

17   Emotions Under the Dome

Helen Ahner

The first planetarium presentations, which took place on the roof of the Zeiss company in Jena in the summer of 1924, triggered a wave of planetarium euphoria. Press reports about the new astronomical science show were full of great emotion. The reporters—above all the Danish astronomer Elis Strömgren—were unanimous: the celestial machine and its projection of the starry sky were true wonders and succeeded in leaving their audience in awe.

The planetarium projector was not the only machine to be spectacularly promoted as a "wonder of technology" in the 1920s. This type of storytelling had been established since the dawn of the twentieth century. New technologies, vehicles and buildings were introduced everywhere as "wonders of technology", then elucidated and made tangible.[14] The popularization of science, with its presentation methods and staging strategies, also often aimed to create wonder and combine the transfer of knowledge with an exciting experience.[15]

Wonder as an epistemic emotion frames a transgressive experience: it occurs when the limits of what was previously knowable, what was considered possible and conceivable are exceeded.[16] Likewise, the sheer magnitude of gigantic (natural) views that challenge and transcend sensory perception is considered such a transgression and causes amazement. People learned what wonder feels like and when it is appropriate to be amazed, for example, in museums, at so-called wonder shows, a mix of science and entertainment, in monumental (church) buildings, when mountain climbing or when visiting tourist attractions that were imbued with an aura of awe.

Newspaper reports from the planetarium in the 1920s also fueled the expectation of being filled with wonder and amazement in the planetarium and prepared the audience for what they would experience under the dome. Emotions do not simply arise in a vacuum, but are cultivated, practiced, socially made and mutable.[17] The framing of the planetarium as a "wonder of technology" alone contributed to people being amazed in the planetarium. In addition, a variety of techniques of wonder were used under the dome, which also contributed to the experience of wonder and awe.[18]

---

[14] Rieger (2003), p 152–176.
[15] Daston and Park (2002).
[16] Geppert (2011).
[17] Frevert (2011).
[18] Ahner (2023).

Fig. 12 Visitors view the new planetarium at Stockholm (BL 06798, Zeiss Archives). ©
Stockholm

The purposeful use of light and darkness, the narrative framing of the projection apparatus, the dramaturgical staging of the "switching on" of the starry
sky and the planetarium furniture, which aligned the bodies of the visitors in
such a way that they were easily amazed with their eyes and mouths open,
helped to make the planetarium guests wonder. The view into the infinity of
space, which as a gigantic natural vista was also considered a trigger for sublime feelings beyond the planetarium (particularly prominent in Kant, for
example), was given a special emotional impact in combination with the
impressive, promising planetarium technology. Even if, considering the technological developments of the last 100 years, a successful projection of the
starry sky no longer seems quite so wonderful, the questions about the secrets
of the universe dealt with in the planetarium remain a source of wonder and
amazement to this day (Fig. 12).

18   Planetarium Jokes

Katie Boyce-Jacino

The year is 1931, and a young man's father has come to visit his son in Berlin
for Christmas. The son, despairing of an activity that both he and his father

will enjoy in the damp and dark of late December, takes his father to the Zeiss Planetarium in the Zoologischer Garten. The man running the planetarium projector is affable and enthusiastic and offers to perform a little trick for the father-son duo. "What time were you born?" he asks the son. "I can show you precisely how the sky looked that night."[19] The son tells him he was born on 13 March 1896, and the planetarium worker quickly sets the machine to spinning, moving swiftly back through the years. The machine settles, and the son is agape, marveling at the incredible sparkling sky that suddenly appears overhead. His father, however, is less pleased. "A racket!" he scoffs, "An absolute, nasty scam! It rained so hard that night, you couldn't see a single star at all!"[20]

This joke appeared in the weekly humor magazine *Ulk* under the title "ZEISS TELLS A FIB," and the modern historian can find it carefully preserved in a folder of newspaper errata labeled "Planetarium Humor" at the Carl Zeiss Optical Company headquarters in Jena, Germany. We find other planetarium jokes in the same collection. In one cartoon from 1935, an old woman charges ten cents to a gullible young couple one evening, promising them an "Open Air Planetarium"—all they have to do after paying, she says, is look up.[21]

Another common motif is that you can get a planetarium show for free simply by getting concussed—a man runs into a lamppost and exclaims "I didn't realize I had already arrived in the planetarium!" as stars spring up around his head.[22] In another cartoon, a man is the unwitting victim of a mechanical malfunction: he already sits in the planetarium, patiently waiting the start of a show, when the heavy dumbbell-shaped planetarium projector accidentally swings down right on top of his head, and stars explode around him. "Here's a way to see the stars in the planetarium!" the caption explains.[23] In another, one man punches another soundly in the head outside a Zeiss Planetarium, causing the usual stars to erupt, while helpfully explaining, "don't get me wrong dear friend! This way I'm saving you the price of admission!"[24]

These jokes appear over and over again in popular references to the planetarium and not just in Weimar Germany; everywhere a planetarium pops up, jokes about seeing stars are sure to follow. Sometimes the stars are concussion

---

[19] ASTRO 0288, Carl Zeiss Archives.
[20] ASTRO 0288, Carl Zeiss Archives.
[21] (2 May 1931) *Zeiss-Werkzeitung* 6.
[22] (2 May 1931) *Zeiss-Werkzeitung* 6.
[23] (2 May 1931) *Zeiss-Werkzeitung* 6.
[24] (2 May 1931) *Zeiss-Werkzeitung* 6.

symptoms and sometimes they're the real stars themselves, but the core of the joke is nearly always the same: the planetarium offers an experience that could just as easily be had elsewhere.

So what can we make of these jokes when considering the history of the planetarium? They tell us two things: first, that the planetarium, whenever it appears, retains a sense of novelty. The notion of a machine built to capture the night sky feels essentially whimsical. These jokes challenge the planetarium's promise to offer an immersive experience of the sky by pointing out that stars (whether cosmic or concussive) already exist for the viewing.

The second conclusion we can draw from these jokes is that despite the skepticism embedded in them, the joke writers nonetheless recognize the lure of the planetarium. They are drawn, reluctantly perhaps, to the promise of it. The planetarium compels engagement—it pulls people into its orbit. These jokes are full of skepticism, but also delight. The idea of such a machine fills even the cynical comic with wonder.

19   Van Gogh's Nocturnal Way Home

Andreas Schmidt

During his lifetime, with only little approval from his peers, but today he is considered by many to be one of the most important and groundbreaking artists of the nineteenth century: Vincent Van Gogh. His expressive emotional painting style anticipated much of what would later become recognized as the principles of modern painting. His life as an outsider and eccentric epitomizes the misunderstood artist, which consequently led to a premature and tragic death. A story like in Hollywood. Somewhat less dramatic will probably be the truth, after all Vincent experienced support from his brother Theo. There is also evidence of a lively exchange with leading Parisian artists of his time. Among others, he made the acquaintance of numerous other painters, including Henri de Toulouse-Lautrec, Paul Signac and Paul Gauguin.[25]

In 1888, Vincent van Gogh moved from Paris to Arles in southern France in search of "another light". Apparently, he found it, for the time in Arles was the most productive of his entire artistic life. One of his paintings created in September of the same year is "Starry Night over the Rhône". It shows the night view on the banks of the Rhône, which was only a minute or two's walk from the Yellow House in Place Lamartine, which Van Gogh rented at the

---

[25] Van Geusau (2016) and Boime (1985)

time. Above the distant lights of the city on the opposite headland and its wonderfully playful reflections in the water, the familiar constellation of the Great Dipper towers clearly visible. It stands erect and is situated just above the horizon. This view corresponds to the view to the northern sky as it can be seen in the mid of September around midnight. A view as it can be easily reproduced in planetariums even today and can be admired again and again.

Maybe once in a lifetime you should travel to the south of France in late summer to experience the special light and walk in the footsteps of van Gogh. Who then stands, after a fine meal and a bottle of wine at night, in Arles on the banks of the Rhône and enjoys the view towards the lights of the city, will be surprised. Where van Gogh placed the Great Dipper, the astronomically trained eye will find the constellation of Aquarius. That is because here you look towards the south, not to the north. Most of us only know Aquarius in its form as a zodiac sign from the calendar. In any case, it is not very impressive in the night sky, which is probably why van Gogh also preferred to simply turn around to paint what is also so conspicuous to him (Fig. 13).

Fig. 13 Vincent van Gogh, Starry Night over the Rhône ("La Nuit étoilée", 1888). The image shows a reproduction of the painting by Art Simeon (oil on canvas, 60 × 80 cm, 2024, photo by A. Schmidt)

20   Astronaut Training at Morehead Planetarium

Michael G. Neece

When NASA announced America's first space explorers as *astronauts* (meaning "star sailors") on April 9, 1959, people everywhere breathlessly anticipated the space adventures soon to follow. But how would these explorers get safely home each time?

Throughout the ages, sailors have used bright stars in familiar constellations to steer a ship to safe harbor, so why couldn't these new star sailors do the same? After a visit to the U.S. Army Transportation School Planetarium in October of 1959, astronauts indicated that star identification training would be vital to piloting in space. Days later, Morehead Planetarium director Tony Jenzano received a call to discuss training astronauts in his star theater.

But why Morehead Planetarium? Morehead was the first planetarium built on a university campus—The University of North Carolina in Chapel Hill, NC. This meant astronomy professors would be close at hand to help develop training with planetarium education staff. And John Motley Morehead III had spared no expense when building the planetarium, so it had a Zeiss II star projector, cutting-edge technology of that era.

Another appealing feature was that Chapel Hill was near the astronauts' flight paths from Langley, Virginia to Cape Canaveral and Houston. It was also a tiny town, which meant less traffic and fewer autograph-seekers.

Jenzano decided astronauts' visits would be announced after each visit, so astronauts had virtual anonymity while in town. Trainers even used code words, referring to astronaut visits as "cookie time" since trays of cookies were laid out when astronauts were soon to arrive. Tony and his wife, Jay, opened their home to the astronauts, befriending several of them, including Neil Armstrong, Deke Slayton, and Gus Grissom. Sixty-two astronauts trained at Morehead from 1960 to 1975 for their Mercury, Gemini, Apollo, Apollo-Soyuz, and Skylab missions. Neil Armstrong trained more than any other astronaut: thirteen visits with more than 130 hours of training.

What did the astronauts learn? Maneuvering a spacecraft involves engine burns during which the spacecraft must point in a particular direction. Stars, being incredibly far away, can act as reference points, like distant landmarks on a road trip. Before any engine burn astronauts would rotate the spacecraft until particular stars appeared outside the spacecraft windows, ensuring proper spacecraft orientation.

This knowledge was used on every mission but came in especially handy during emergencies. On the *Faith Seven* Mercury flight in 1963, electrical systems malfunctioned, so Gordon Cooper steered his ship manually while watching the stars outside his window, piloting his spacecraft to the most accurate splashdown of any Mercury flight. When Apollo 12 was struck by lightning during liftoff, Dick Gordon had to restart and realign the navigation system using his knowledge of the stars. These two instances demonstrated the power of Morehead training.

Most do not know the story of Chapel Hill being the secret space town in the 1960s and 70 s because it was quiet by design, but there is no doubt that America's first astronauts made it home safely because of their time at Morehead Planetarium.

21   Space Travel Shows at the Griffith Observatory in the 1950s

David DeVorkin

Fanciful "Travel to the Moon" lectures and demonstrations have been appearing since the late nineteenth century using magic lanterns variously called Sciopticons, as Frank Winter has described in his detailed history on the subject.[26] Starting in the 1920s full dome optical planetariums served primarily as portals to the heavens and to the Earth's place in the visible universe. The Hayden Planetarium in New York City and the Fels in Philadelphia initiated fanciful spaceflight shows in 1938 and 1939.[27] But by the mid-to-late 1950s, planetaria were producing more realistic journeys into space, to the Moon and the planets, employing the physically based yet wonderfully romantic space art by visionaries like Howard Russell Butler and Chesley Bonestell from the 1940s that the artists modified to produce panoramic 360-degree horizons of the Moon and planets against a sea of stars. This essay briefly reviews the known history as cited above.

The Hayden and Fels initiative may well have stimulated other planetarium directors like the noted selenographer Dinsmore Alter of the Griffith Observatory, who in 1948–1949 produced a series "of new type shows" utilizing sophisticated optical zoom systems called "space travel projectors" designed

---

[26] Winter (2009), p 12–13.
[27] Marché (2005), p 75–76.

and built by his staff in the observatory shops.[28] He also used a live action overhead transparency projector in his shows to project large positive transparencies of celestial scenery that would allow him to manually zoom to their vicinity and examine them in detail. The first series in 1949 included "A Visit to the Frozen Planets, Jupiter and Saturn," and next, in the summer, came "A Trip to the Moon." In the fall "A Visit to the Sun" was followed by "Roaming the Milky Way." Alter admitted that the latest show "is far beyond any distance to which, at present, we can hope future development to take mankind." The Griffith staff used the best earth-based telescopic views from Mount Wilson and Lick Observatories to simulate these voyages. At first, these shows used only telescopic views.

In 1957, Alter reminisced that even before World War II, he felt that "the preparation of the public for possible future space travel was accepted as one phase of the mission." He had raised the question of "Observatories on the Moon?" in the September 1940 issue of the Griffith Observer and raised it again later in the 1940s and especially in 1952 but this time without the "?."

In "A Trip to the Moon" Alter first transported the audience in a suborbital flight to an equatorial island, seen around the horizon of the dome projected by a series of wide-angle projectors featuring a great space ship on its launch pad. Entering the ship, control surfaces would be projected on the screen, the rockets would roar, fade away, and eventually the Moon was reached. Alter would then slide the transparencies across the surface of the overhead projector to cruise the valleys, maria and cratered landscape, pointing out known and interesting features; features he had studied personally through the eyepiece of a great telescope, a world he knew intimately as a long-time student of the Moon.

The show did not reach the Moon's surface until 1952 when it was retitled "We Land on the Moon" with a dramatic set-down in the crater Copernicus. Visitors were treated to a dramatic panorama created by a series of eight overlapping photographs of the interior crater wall of a plaster casting of 12-foot a model of Copernicus for the observatory by the Hollywood model maker Leon Bayard DeVolo. The model was very costly, "but it is believed that the new spectacle fully justifies the expenditure."

In 1957, as a promotional for his new book *The Conquest of Space* with Willy Ley, Bonestell loaned the Observatory a set of 31 astronomical paintings of the Moon and planets for a limited time for display. A year later Bonestell paintings of a hypothetical lunar landscape started appearing in the shows. And by 1959, the show was further enhanced by a landing on Titan,

---

[28] Alter (1949), Bunton (1951).

again illuminated by Bonestell's art. In his teenage years, the author keenly recalls Alter, the elderly white-haired selenographer, land the ship, letting Bonestell's panoramas take over, leaving him quite breathless.

22   Competing Projectors

Matthew McMahon

By the 1960's there was a competitive market for those looking to purchase a planetarium projector for their institution. Armand Spitz has been detailed elsewhere in this volume (Armand Spitz, Ben Buhl), as have many of the other major companies active. By 1964 an aspiring planetarian director, with sufficient financial backing from their government, university or board, could choose from Spitz Laboratories, GOTO, Minolta Camera Company, and Zeiss all offered powerful opto-mechanical projectors to suit a variety of budgets.

Armand Spitz had first become involved with Dr Eric M. Lindsay, the Director of the Armagh Observatory, in the 1940's. He offered firm support to Dr Lindsay as he planned to build a planetarium in Armagh to accompany his historic observatory. Over the next two decades there would be numerous disappointments as the funding failed to materialize due to a sluggish economy, and a lack of enthusiasm at the highest levels of government. However by 1965 the funding had been secured, and both men were keen to see a Spitz A-3-P Projector at the heart of the new institution.

Patrick Moore, who had come to international attention due to his popular astronomy television program on the British Broadcasting Corporation, was appointed Director of the new institution and he visited the United States to see Spitz Projectors in action. By this time Armand had retired from the company, but he kept in contact with Dr Lindsay. However in March of that year the GOTO Optical Company threw their entry in the running, and demonstrated their 'Mars' or Model M-1 projector to Patrick Moore. Eventually it was decided to proceed with a contract for the GOTO instrument, and Spitz was notified that the A-3-P would not be purchased.[29]

Armand Spitz wrote a short letter to Dr Lindsay, which in many ways set the tone for the planetarium sector. He was evidently disappointed but he also wished to emphasis his enthusiasm for the planetarium projector, and the world of Planetaria (Fig. 14). The following is an extract:

---

[29] McMahon and Nežič (2024), p 37–43.

ARMAND N. SPITZ
500 BARKLEY DRIVE
MANTUA, FAIRFAX
VIRGINIA

280-5216
AREA CODE 703-2XXXXX

Dear Eric:

It was distressing.... but I must confess, not
surprising.... to hear that Goto is beginning to be
so much more impressive to your committee. I know of
so many cases in which their bid for acceptance has been
made ~~their ers~~ so forcefully that the entire selection
process has become unduly drawn out and has even resulted
in unpleasantness. This is not the background which
should be a part of planning for any planetarium installation. Maybe I'm an incurable idealist, but methinks that
planetariums and all they stand for should bring people
together and should not be separative in effect.

Any planetarium, Goto, Zeiss, Spitz or any other,
that is well conceived and operated will do an acceptable
job. It's all XX dependent upon the spirit of its
sponsors and operators. I would rather not take any active
part in the sales effort on behalf of Spitz.... as you know,
I'm pretty thoroughly removed from it. I must, in all
honesty, say that I have met several people recently who,
having decided upon Goto instead of Spitz, expressed their
regret about it. Only yesterday I heard that Joel Martin,
who had given every indication of deciding on a Goto, had
bought a Spitz. So, very frankly, I don't know what to
say, and I'd rather not be placed in the position of
trying to influence you in any way.

One thing I emphasize to you... and I told this
to Patrick when he was here: I believe in planetariums
as an educational and cultural asset to any community.
I have offered to lend him and you any support within my
power whatever instrument you buy, and I mean it. Naturally I'd be happy if it bore my name, but my interest in
your success is not predicated upon this. Dost understand?
There is no reason why Goto cannot make a good instrument. Just caution your committee to weighX all things
carefully in making their decision, and call on me if you
feel that I can help.

My best to Pat when you see him. Grace and I had a
nice note from Father McCarthy yesterday. We missed him
when he was here for his last treatments.

Good luck.... and I'll be interested in hearing
whatever happens.

As ever,

Fig. 14 Letter to Dr Lindsay from Armand Spitz, 1965 (Armagh Observatory and
Planetarium)

…Maybe I'm an incurable idealist, but methinks that planetariums and all they stand for should bring people together and should not be separative in effect.

Any planetarium, Goto, Zeiss, Spitz or any other, that is well conceived and operated will do an acceptable job. Its all dependent upon the spirit of its sponsors and operators.

One thing I emphasize to you…and I told this to Patrick when he was here: I believe in planetariums as an educational and cultural asset to any community.

References

Ahner H (2023) Planetarien: Wunder der Technik - Techniken des Wunderns, 2nd edn. Wallstein Verlag, Göttingen

Alter D (1949) Roaming the Milky Way. The Griffith Observer

American Institute of Physics (2015) https://www.aip.org/history-programs/niels-bohr-library/oral-histories/32485

Bolt M (2009) Telescopes: through the looking glass. Adler Planetarium & Astronomy, Chicago, IL

Bunton G (1951) A universe indoors. ASP Leaflet No. 262

Buttles LI (1932) Amateur telescope makers in Chicago. Popular Astronomy 42:163

Callum WM (1936) Amateur telescope makers of Chicago. Popular Astronomy 44:103

Cameron G (2010) Public skies: telescopes and the popularization of astronomy in the twentieth century. PhD thesis. Iowa State University

Daston L, Park K (2002) Wunder und die Ordnung der Natur. Eichborn, Berlin

DeGaff J, Hamill F (1972) Seeing stars. The Planetarian 1(2)

Fox P (1935) Adler planetarium and astronomical museum. An account of the optical planetarium and a brief guide to the museum. The Lakeside Press, R. R. Donnelley & Sons Co., Chicago, IL

Frevert U (2011) Emotions in history – lost and found. Central European University Press, Budapest

Ganz CR (2008) The 1933 Chicago world's fair: a century of progress. University of Illinois Press, Urbana, IL

Geppert ACT (2011) Wunder: Poetik und Politik des Staunens im 20, 1st edn, Jahrhundert. Suhrkamp Verlag Ag, Berlin

Havlik RJ (2006) A fair use of Arcturus: a syzygy of scholarians and the lighting of the Chicago century of progress exposition, 1933-1934. J Astron Hist Herit 9(1):99–108

Korey M (2012) Transatlantic inspiration: the Mathematisch-Physikalischer salon in Dresden and the founding of the Adler Planetarium in Chicago. In: Bolt M, Case S (eds) Engaging the heavens: inspiration of astronomical phenomena (Astronomical Society of the Pacific). Astronomical Society of the Pacific, San Francisco, CA

Marché J (2005) Theaters of time and space: American Planetaria, 1930–1970. Rutgers University Press, New Brunswick

McCray WP (2008) Keep watching the skies! The story of operation Moonwatch and the Dawn of the space age. Princeton University Press, Princeton

McMahon M and Nežič R (2024) Archival research in a planetarium: The first projector at Armagh Planetarium. Communicating Astronomy with the Public Journal (34): 37–43.

Osterbrock DE (1997) Yerkes observatory, 1892–1950 – the birth, near death, & resurrection of a scientific research institution: the birth, near death and resurrection of a scientific research institution. University of Chicago Press, Chicago, IL

Raposo P, Graney C (forthcoming) "A true and exact description of the sun's palace": constructing the image of the solar system (c. 1660-c. 1750) in J Astron Hist Herit

Rieger B (2003) 'Modern wonders': technological innovation and public ambivalence in Britain and Germany, 1890s to 1933. Hist Work J 55

Spitz A (1972) Armand Spitz at CAPE, 1970. The Planetarian 1(1)

Taub L (1995) 'Canned astronomy' versus cultural credibility: the acquisition of the Mensing collection by the Adler Planetarium. J Hist Collect 7(2):243–250

Williams T (2000) Getting organized: a history of amateur astronomy in the United States. PhD thesis. Rice University

Winter F (2009) To the moon by Sciopticon. The Griffith Observer 73

Zeiss-Werkzeitung 6 (1931) Das Planetarium im Humor des In- und Auslands

People in the Planetarium

Ben Buhl, Yaël Nazé, Aubrey Henrietty, Michelle Nichols,
Volkar Schorcht, Pedro M. P. Raposo, David DeVorkin,
Andrew Johnston, Katie Boyce-Jacino, Mike Smail,
Nigel Henbest, Stephanie Ridley, Noreen Grice, Hachioji,
Wolfgang Steffen, Nico Koning, Kerem Osman Çubuk,
Stefano Giovanardi, Chris Helms, Arjun Chawla,
and Ka Chun Yu

1   Introduction

Planetaria are not autonomous systems that, when left to their own devices, can function. The projectors, buildings and exhibitions require people to perform and people to view their performances. The people featured in this chapter range from the titanic figures who have left an undeniable mark on the sector, to those everyday planetarians still working in the archives, at the front desk, and behind the projector. This chapter is also looking at how the visitors

_____

B. Buhl
Exton, PA, USA

Y. Nazé
University of Liège, Liège, Belgium

A. Henrietty • M. Nichols • K. Boyce-Jacino • M. Smail • C. Helms
Adler Planetarium, Chicago, IL, USA

V. Schorcht
Carl Zeiss Jena Planetarium Division (retired), Jena, Germany

P. M. P. Raposo (✉)
Academy of Natural Sciences of Drexel University, Philadelphia, PA, USA
e-mail: pmr64@drexel.edu

© The Author(s), under exclusive license to Springer Nature Switzerland AG 2024   **145**
M. McMahon et al. (eds.), *100 Years of Planetaria*, Springer Praxis Books,
https://doi.org/10.1007/978-3-031-75496-8_4

interact with the planetarium, and how the planetarium can interact with the local community.

This chapter also reflects the point in time we are currently at, one hundred years since the first projector was opened to the public in the early 1920's. Some of the people featured in this chapter have dedicated themselves to preserving the history of the sector, through institutional archives, instruments and research. It also looks forward, with some sharing their hopes and ambitions for the world of planetaria in the next century, as well as ways in which people can use the new technology to better understand their universe.

D. DeVorkin
Smithsonian's National Air and Space Museum, Washington, DC, USA

A. Johnston
Adler Planetarium, Chicago, IL, USA

National Air and Space Museum, Washington, DC, USA

N. Henbest
Hampstead, NC, USA

S. Ridley
Morehead Planetarium and Science Center, Chapel Hill, NC, USA

N. Grice
You Can Do Astronomy LLC, New Britain, CT, USA

Hachioji
Tokyo, Japan

W. Steffen • N. Koning
Ilumbra - AstroPhysical MediaStudio, Kaiserslautern, Germany

K. O. Çubuk • A. Chawla
Armagh Observatory and Planetarium, Armagh, UK

S. Giovanardi
Planetarium and Astronomical Museum of Rome, Rome, Italy

K. C. Yu
Denver Museum of Nature & Science, Denver, CO, USA

2   Armand Spitz

Ben Buhl

Planetarium innovator, educator, and advocate extraordinaire Armand Neustadter Sptiz (1904–1971) stands today as one of the industry's legendary visionaries. His steadfast passion and forward-thinking nature exemplify the Centennial of the Planetarium and its mission to exponentially expand world-wide planetarium awareness, especially with respect to the field's growing impact on science, education, technology, the arts, and math.

In 1947, there were only a few planetariums in the United States. Armand Spitz was a lecturer at one of them—the Fels Planetarium in Philadelphia. The position had taken him several years to secure, yet he was not content knowing most of the world lacked access to this incredible learning resource, mainly due to various prohibitive costs.

No barrier too high, Spitz was so convinced of the planetarium's educational potential that—despite having little funds and no formal science or engineering degree—he designed his own projector, the Model A, thanks in part to an assist from fellow astronomy enthusiast Albert Einstein. In 1947 the projector sold for approximately $7,000 USD in today's money—a monumental reduction in cost from all others available. Over the next ten years the groundbreaking innovation was followed by several advanced models as the number of planetariums rose from five to more than 200.

Another significant Spitz contribution to planetarium history came in the shape of the "Spitz Junior Planetarium." Sold in a shoebox-size package with eye-catching graphics and an accompanying illustrated booklet, the "pinhole-style" planetarium saw over one million units produced between 1955 and 1972 at a modern-day purchase price of about $150. Although sometimes referred to as a toy, the Spitz Junior was actually a high-quality, sophisticated instrument—less a plaything and more a device for anyone, including educators, seeking a unique and exciting way to become better acquainted with the nighttime sky. Today the item is a sought-after vintage collectible available through various vendors including the Smithsonian National Museum of American History.

Since departing Earth's celestial sphere in 1971 at the age of 66, Armand Spitz retains a legacy that continues to grow. The company he founded in 1947 in a small Pennsylvania (USA) town is now the worldwide leader in projection domes with over 1,200 installations. Beyond measurable achievements, his life is a model of passion, perseverance, and desire emblematic of the Centennial of the Planetarium's global campaign to make known the planetarium's boundless imaginative and practical possibilities.

3   Reysa Bernson

Yaël Nazé

France was quite late to host planetaria as the first one was only installed in the country in 1937. This occurred as part of the World Exhibition organized in Paris that year. This exhibition actually had three astronomically-related features: the astronomy display in the prestigious "Palais de la Découverte" (imagined and managed by professional astronomers of Paris observatory) and two showcases in the nearby amusement park, a mock rocket travel called "stellarium" and a planetarium. In fact, adding a planetarium was considered quite early for the Paris Exhibition (see notably a mention in a 1934 mail exchange between the French embassy in Berlin and Paris' Luxembourg museum, Bergeron & Bigg 2015),[1] probably because of the successes of the previous exhibitions' planetaria in Chicago and Brussels. However, the Palais organizing committee, composed of professionals, soon discarded the idea. It is therefore unclear how and when the idea surfaced again but, in 1935, the contract was signed and the planetarium started to be repeatedly mentioned in newspapers, being presented as a highlight of the coming Exhibition.

Unsurprisingly, the team chosen to present the planetarium shows was not composed of professional astronomers, but of amateur astronomers: the SAF secretary André Hamon, the autodidact Eliezer Fournier, the astrophotographer Henri Kanapell, sky observers Jacques Codry and Auguste Budry, and the young students Armand Delsemme and Gérard Oriano (later Gérard de Vaucouleurs). The head of this team was Reysa Bernson (1904–1944). Her name has been nearly totally forgotten but she was one of the most important popularizers of the time.[2] Nevertheless, the choice of a woman, and of this woman in particular, can appear puzzling considering the epoch. The exact path that led to this decision remains unknown but she had promoted planetaria since she saw one in Dresden in 1931 on a students' trip and there were links between one of the Exhibition organizers (A. Léveillé) and her.

Reysa Bernson was born from two M.D., her father was French but her mother came from Belarus. She took her mother's name after the separation of her parents soon after WWI. She went to Lille university where she gathered a bachelor and a master in Sciences, along with several certificates including one on astronomy. She became interested in astronomy at a young age (she mentioned the hybrid eclipse of 1912 in an interview and became a member of the "Société Astronomique de France" in 1920, at only 16 years old). She

[1] Bergeron, A. and Bigg, C. (2015).
[2] Nazé, Y., (2023), p 816–832.

founded Lille's "Astronomical Association of the North" in 1923 and unsuccessfully tried to create an international organization for amateurs, to complement IAU. As amateur astronomer, she performed a lot of observations, contributing more than 4000 points in the AFOEV database of variable stars' observations. In this context, she proposed in 1935 to separate novae in several families, as did nearly simultaneously the astronomers Lundmark, Gerasimovic, and McLaughlin, who are generally credited for the discovery. In parallel, she developed an intense outreach activity, with dozens of public talks, radio or newspaper stories, school and scout workshops. Her exceptional skills in this domain were broadly recognized, as testify media comments and the granting of several prizes. She never hesitated to use the latest technical possibilities (e.g. movies, radio) for her outreach and she had a very modern view of how it should be done (e.g. inclusion of disabled students, active learning, lively speech…). However, she had to repeatedly face gender and/or race prejudice. WWII put a definitive stop to her career. Because of the Jewish origin of her mother, she had to hide and could not continue her astronomical activities. She was arrested in 1944, deported and killed in Auschwitz. The 1937 planetarium also had a sad ending: dismantled at the end of the Paris exhibition and then sold in 1938 to the city of Paris, it remained stored in the cellars of the CNAM until 1952… when it was installed in the Palais which originally did not want it!

4   Maude Bennott

Aubrey Henrietty

Before Maude Bennot became the second director of the Adler Planetarium in 1937, she was already living a pretty extraordinary life. She had an advanced degree in astronomy. She was an amateur violinist and pilot. She was smart, ambitious, ahead of her time.

Unfortunately for Maude, her time was the first half of the twentieth century and she was a woman, which was even more of a personal and professional liability then than it is now, a hundred years later.

While she filled her days running the theater, giving lectures, writing budgets, paying bills, answering the phone, lending her expertise to help solve an astronomy clue in a crossword puzzle, or training military pilots to navigate by starlight at the Western Hemisphere's first planetarium, many powerful people were convinced that she didn't belong there. Colleagues, journalists, and city officials fussed over her gender, physical appearance, and marital status.

Even the ones who seemed to be on her side—like the writer Sydney J. Harris—couldn't resist repeating some of the sexist and condescending things other people had said about her before assuring readers that those people were wrong.

"Park District officials, who operate the museum, were skeptical of this slim, fragile woman," Harris wrote in the Chicago Daily News. "Masculine astronomers shook their heads dolefully, said she was more in place in a tearoom than in an observatory. Today, they have chewed those words into very tiny bits."

They didn't have to chew on their words forever. In 1945, the Chicago Park District fired Maude—allegedly due to her "insubordination," a charge that may have been related to her efforts to keep political appointees off of her staff. But rumors had been circulating at least since the previous year that the officials had been angling to replace her with a man. If Maude's tenure at the Adler Planetarium had ended differently—if she had retired at a time of her own choosing instead of being fired and replaced with a less-qualified man— we would remember her as the person who kept America's first planetarium alive through a great depression and a world war.

We would remember how she took to the airwaves on Chicago's WBBM radio, where she hosted a show about astronomy for six months in 1940, when the planetarium was closed for repairs. We would remember that, as far as we know, she was the first woman in the world to run a science museum. But the circumstances of her firing force us also to remember how hard she had to fight for the privilege of serving the public in those ways—and how hard the systems of power and authority that felt threatened by her fought back.

5   Henry C. King

Matthew McMahon

Henry C. King was born 9 March 1915, in London. His interest in astronomy was lit by a copy of the *The Story of the Heavens* by Astronomer Royal of Ireland, Robert Ball, which he received for his thirteenth birthday. Though he did not attend university, he completed a bachelor's degree in astronomy by correspondence with the University of London. He would go on to receive his Masters and then a PhD in 1951 (on which he would base his first book), and during the Second World War he was an inspector at the Ministry of Aircraft

Production. He would go on to become the first and only Scientific Director at the newly built London Planetarium in 1956.[3]

In 1966 he was invited to become the Director of another new planetarium, the McClaughlin Planetarium in Toronto, Canada. He accepted and joined the institution, which was linked to the Royal Ontario Museum in conjunction with the University of Toronto. The project was funded in part by a local businessman, Colonel Samuel McClaughlin, and the building was designed by George Gouinlock and Hugh Allward. As Director, Henry King opened the planetarium to the public on 26 October, 1968.

He would go on to become one of the finest historians of astronomy in the twentieth century, producing two exceptional textbooks which have become standard reference books in many observatory and planetarium libraries all over the world. *The History of Telescope* was published in 1955 and was a journey through the history of the optical telescope to the (as it was in 1950's) present day.[4] It was an elaboration on his 1951 doctoral thesis, and had come at an important period, during which the history of astronomical instrumentation was undergoing a revival. Very few telescope makers escaped his pen, and it still provides an unparalleled source of information gleaned from archives, trade catalogues and scientific journals all over the world. *Geared to the stars* came in 1978, written in conjunction with John R. Millburn, with whom he maintained a long correspondence and collaborated on further books with.[5] This was greatly influenced by his time not only researching astronomical instruments, but also working as a planetarium director. He drew a long historical arc, from the earliest astronomical clocks and orreries, to the modern projection planetarium, and also paid attention to those who operated the instrument, taking care to explain exactly how different types of dome shows were delivered to the public in different institutions across the world.

6   Caretakers of the Sky

Michelle Nichols

Over the centuries, humans have built many different types of machines to represent and recreate the sky, celestial objects, and their movements. On 5 June 1913, a new exhibit debuted at the Chicago Academy of Sciences (CAS) museum: the Atwood Celestial Sphere. The Sphere is a mechanical

---

[3] King DA (2007), p 526–527.
[4] King HC (1979).
[5] King and Millburn (1978).

planetarium, originally designed by geologist and CAS Board member Dr. Wallace Atwood. The structure consists of a large ball that is approximately five meters (15 feet) in diameter and skinned in thin galvanized sheet metal. Light shining through 692 small holes in the metal sheeting approximates the locations of many stars down to 4th and 5th magnitude that are in the sky at Chicago's 42-degree north latitude. A program facilitator operated a motor to rotate the sphere, showing the small audience inside how the stars appear to move, and the earliest iterations of the interior also included moveable depictions of the Sun, Moon, and planets. This remarkable sky globe served CAS audiences for decades via exhibits surrounding it and facilitated programming inside it.

In 1997, the Sphere made its way to the Adler Planetarium, the western hemisphere's oldest projection planetarium facility. It went on display to the public starting in January 1999. Since that time, Adler's guests have been able to see and experience mechanical and projection planetariums just steps from each other. The Atwood Sphere is currently the world's oldest functioning sky simulator, operating alongside some of the world's most advanced digital projection systems in the Adler's two large domed planetarium theaters.

At various times, the Sphere has been in need of a little upkeep and care. I have had the opportunity to get my hands onto the Sphere's interior several times to give it periodic facelifts. Completing the refurbishment work has included inserting (and, subsequently, removing) hundreds of toothpicks to fill the star holes as I sprayed the Sphere's interior with gallons of white primer and black paint, brushing dozens of feet of thin lines of UV-sensitive paint, a few short strokes at a time, to highlight the outlines of constellations, and hand-numbering all 692 star holes, not once but twice. When I look closely, I can still see and feel traces of prior lines and numbers and other long-ago details of the Sphere's interior barely visible under the layers of paint. When the Sphere turns, the sheet metal flexes and rumbles: it "speaks" to us. Working so meticulously on this object has become a tangible connection from me to those who came before me to create it, use it, and maintain it, all the way back to Dr. Atwood himself. It is not lost on me how incredibly unique this experience is. My Adler exhibit maintenance and history of astronomy colleagues and I are fortunate to be the latest in the line of people who have attended to Dr. Atwood's marvel over the years.

The Adler Planetarium carries on Dr. Atwood's goal of using the Sphere to educate and inspire the public. I sometimes wonder what he might have thought about it being used today—or if he ever thought it would still exist more than a century later. I hope our continued care of the Sphere will allow the Adler to use it to educate and inspire for another century—and I hope Dr. Atwood would be proud of our efforts.

7   Bridges over the Wall

Volkmar Schorcht

For forty years, the Wall divided Germany into East and West. There were planetariums on both sides, and not only that! The Carl Zeiss company, which was also separated when Germany was divided, developed and produced planetariums in the West (Carl Zeiss Oberkochen) as well as in the East (Carl Zeiss Jena). Both companies competed and installed planetarium projectors all over the world.

For decades, ideological politics in the East prevented cooperation between the planetariums of the two German states. It was only in the 1980s that the Wall could be bridged in a few places. As a result of a compensation deal between the Volkswagen Group—VW delivered 10,000 Golf cars to the East—and the German Democratic Republic, the city of Wolfsburg in the West received a planetarium from Carl Zeiss in Jena from the East. It opened in 1983. Volkmar Schorcht, who took over the management of the Jena Planetarium in 1985, came from the Carl Zeiss company in Jena and was familiar with the special business relationship between Jena and Wolfsburg. When the director of the Wolfsburg planetarium, Dr. Bernd Loibl, put forward the idea of also presenting special events conceived by him in the Jena planetarium, Schorcht succeeded in setting the appropriate course in Jena. From then on, numerous guest performances of the Wolfsburg Planetarium could take place in Jena. The focus was on the Hamburg actor Rudolf H. Herget, who performed literary texts live in the planetarium dome, while Loibl and the Jena planetarium team with Frank-Michael Arndt, Werner Don Eck and Dr. Hans Meinl provided the audiovisuals. Herget recited classical poems, quoted from Antoine de Saint-Exupery's "The Little Prince", Richard Bach's "The Seagull Jonathan" and brought poetry to life. The performances at the Zeiss Planetarium Jena were always sold out. The two planetariums thus built small bridges between East and West, bridges that ultimately helped bring down the Wall in Germany.

8   Eugénio da Conceição Silva

Pedro M. P. Raposo

Eugénio da Conceição Silva (1903–1969) might not be a well-known name outside the Portuguese Navy and the national amateur astronomy community, but in those circles he is celebrated to this day as the founder of the first planetarium in Portugal, the Gulbenkian Planetarium (see *The Gulbenkian Planetarium in Lisbon, Pedro Raposo*), and as an accomplished astrophotographer who continues to inspire new generations of amateurs—all of this adding to a sound record as a naval officer.

After completing his training at the Portuguese Naval Academy, Silva specialized in naval artillery and optics. He nurtured an interest in astronomy from a young age, becoming a member of the Société Astronomique de France while still a cadet. In 1935, he engaged in systematic observations of variable and double stars.[6]

Between 1936 and 1958 Silva produced two photographic atlases of the sky,[7] using different telescopes and cameras. The atlases were never published but Silva's work came to the attention of many fellow amateurs abroad. In September 1952 his private observatory at the Alfeite Naval Base in the Lisbon area was featured in the column 'The Amateur Scientist' in *Scientific American* edited by Albert Ingalls, who was also the editor of Amateur Telescope Making (see *Making Telescopes at the Adler Planetarium, Pedro Raposo*). Realizing that as an amateur astrophotographer, no matter how skilful and resourceful, he would never be able to compete with leading observatories such as Mount Palomar, Silva decided to focus on writing popular books and articles about astronomy, space exploration, and other scientific topics.

A visit to the Hayden Planetarium in New York in the late 1940's seemingly inspired him to promote and lead the foundation of a planetarium in Lisbon, an endeavor that came to fruition with the inauguration of the Gulbenkian Planetarium in 1965, with Silva as the first director. While as a naval officer Silva often took on official commissions from the dictatorial regime then ruling over Portugal, the so-called Estado Novo, he is said to have often expressed his loath of the dictatorship in private. According to an anecdote well known in the circles where Silva is fondly remembered, he compressed the inaugural planetarium show into 10 minutes and fled from the building as soon as he could, just to avoid crossing paths with the President

---

[6] Silva E.C. (1942).
[7] Silva E.C. (1948), Silva E.C. (1958).

and other Estado Novo dignitaries in attendance. Still, he was always keen to extol the historical role of astronomy in the Portuguese maritime past that the regime invoked and celebrated to justify the maintenance of its overseas colonial empire.

At the same time, Silva also promoted optical workshops open to the public, in which participants learned how to grind and polish mirrors and to assemble reflecting telescopes. He also planned to set up an observatory next to the Planetarium, which was only built in 1972, three years after Silva's death, but eventually named after him. Initially equipped with a refracting telescope borrowed from the Faculty of Sciences of Lisbon, it provided another space for engaging visitors with astronomical observing. These activities were pivotal in setting the grounds for the formation of the Portuguese Association of Amateur Astronomers. While Silva compromised with a political regime he allegedly detested, he was successful in steering a planetarium project that played a crucial role in promoting astronomy in Portugal and in consolidating an amateur community in the country.

9   The first person to place an observatory on the Moon made a telescope mirror at the Adler Planetarium - George R. Carruthers

David DeVorkin

In 1972, Naval Research Laboratory astronomer George R. Carruthers sent a small but very powerful dual Schmidt camera/spectrograph to the moon on Apollo 16. Once there astronaut John Young followed Carruthers's directions to image the Earth's outermost atmosphere and to explore stars, nebulae and galaxies in the extreme ultraviolet. It was the first astronomical observatory on the Moon, and to this date the only one.

Since childhood Carruthers was passionate about space travel and astronomy. He built his first small telescope in the late 1940s with lenses found in a magazine advertisement and observed the heavens from his family farm in Ohio. When his mother closed their farm and moved the family to Chicago in the early 1950s Carruthers, a young teen, soon found the Adler Planetarium, which greatly expanded his astronomical universe.

In that day, major planetariums, including the Adler, the Griffith Observatory and the Hayden Planetarium, among others, created workshops and gave courses on astronomical topics.[8] They supported informal sessions and facilities to make telescope mirrors, selling kits including mirror blanks, ceramic tools, and a graded set of carborundum grits to grind the mirrors to the right shape, and then to polish them with rouge to complete their spherical or parabolic figures. Adults of all walks of life had open access to these facilities and children were strongly encouraged to join in. And to be sure, the experience was transformational producing graduates who went on to become noted astronomers.[9]

Living on the south side, in Englewood, Carruthers took the 63rd Street "L" elevated train downtown to the Adler, attending informal events, reading in the library and taking classes. His favorite was a telescope mirror-making class directed by Dr. Albert Shatzel, an assistant director of the planetarium. Carruthers ground and polished a 4–1/4 inch mirror there, all that he could afford. At home, he built a Newtonian telescope with a tube made of wood and a mounting from plumbing parts, adding a small flat glass mirror to complete the Newtonian design. Encouraged by his teachers at Englewood High School, he entered his telescope into a local science fair.

The Adler programs were open to everyone. With free access to the library Carruthers devoured books on space travel and the means to get into space, like Milton Rosen's *The Viking Rocket Story*, published in 1955. He attended informal evening events there too, and mingled with others his age fascinated with astronomy. As he recalled in an interview, "Certainly, the planetarium was an [order of] magnitude jump over anything that I had been exposed to in Milford, so that caused my interest to go up another order of magnitude, you might say."

But there was one memory that Carruthers keenly shared: Although he enjoyed the mentoring of the Adler astronomers, when Carruthers's expressed his fascination linking astronomy and space travel, stimulated by the famous Collier's series of articles, "the astronomers that I talked to at the planetarium thought that was nonsense, that astronomy is done with ground-based telescopes, and you shouldn't waste your time thinking about going out into space."[10] All that changed in a few years with Sputnik I.

---

[8] Raposo P (2020), Williams (2000), Adapted from : de Vorkin D (2025) *From the Laboratory to the Moon: The Quiet Genius of George R. Carruthers*. MIT.

[9] Chilton L (2017), p 6–8.

[10] Physics AIP (2015).

10   Preserving a National and International Planetarium Icon

Andrew Johnston

The National Air and Space Museum opened in 1976 in Washington DC, one of many celebrations for the 200th anniversary of the nation's founding. The new museum included the Albert Einstein Spacearium, later called the Einstein Planetarium. It featured a Zeiss mark VIa projector, a donation from the Federal Republic of Germany. This was the official gift from West Germany to the United States for the American bicentennial. The Zeiss projector was used for 43 years, including many years with a digital display system installed in 2002.

I worked at the National Air and Space Museum until 2016, using the Zeiss for "Stars Tonight" sky lectures and other programs along with several other team members. The Zeiss projector occasionally needed repairs. Working with the exhibits shop, we aligned the planets and made sure gears meshed correctly. One mechanical failure occurred with the Mars projector, requiring fabrication of a new drive shaft. It was literally celestial mechanics.

An essential element in maintaining the projector was the collaborative nature of the planetarium community. Planetarium colleagues generously allowed us to acquire needed spare parts as other Zeiss VI projectors were retired. At the Morehead Planetarium in Chapel Hill, North Carolina, I disassembled planet projectors and removed several gears from their Zeiss VI. From the Fiske Planetarium in Boulder, Colorado I acquired parts from the Zeiss VI formerly used at the Museum of Science, Boston. On a trip to the Adler Planetarium in Chicago, I checked out the Zeiss VI from the American Museum of Natural History in New York. Much of that machine was later transferred to the Strasenburgh Planetarium at the Rochester Museum and Science Center. At the Adler, where I have worked since 2016, we continue to hold parts of Zeiss VI projectors and our original 1930 Zeiss. My luggage on those trips contained specialized tools, large gears, planet projectors, electrical connectors, and occasionally bottles of beer from local breweries. That often attracted attention from security, but everything arrived safe and sound. We installed several of these gears and connectors in the National Air and Space Museum's projector, and they remain in that machine to this day.

Later in the 2010s the National Air and Space Museum began planning a significant renovation that required closing the building in stages. The planetarium would close for installation of a new dome and seats. It was finally time to retire the Zeiss projector. The curatorial team began work to accession the projector into the collection. In 2019, the projector was removed by museum staff and then reassembled for public display. The projector and console can now be seen at the National Air and Space Museum's Hazy Center near Dulles Airport outside Washington DC. It is a fitting place to share this icon of national pride and international cooperation that awed so many people over the years.

11   A Personal Journey in Astronomy Begins

Pedro M. P. Raposo

The year was 1993. I was a teenager growing up in the Portuguese city of Barreiro, in the industrial outskirts of Lisbon. My hometown was known for its chemical industries, the railways that connected the Portuguese capital with the south of the country, and the nearby national steel works. In short, a satellite industrial town that no one wanted to go to unless they lived or had family there. Although the industrial chemical complex was already in a process of decline that eventually led to its deactivation, pollution was still very noticeable and present in the life of the city.

This might not sound like the best setting for a lifelong fascination with the sky to emerge, but that is where it happened to me. In that summer of 1993, astronomy was making headlines in Portuguese television and newspapers. The annual August meteor shower of the Perseids was allegedly going to be spectacular as the comet in whose trail the meteors originate, P/Swift-Tuttle, had made its periodic return to the vicinity of Earth. I would later learn that it is always a good idea to curb expectations about meteor showers, and that a full appreciation of even the most remarkable events of this kind requires good observing conditions and an attentive counting of the number of meteors observed throughout the night. At the time, I got just as excited by the media hype as anyone else, and sure enough I eagerly looked up on that warm August night when the Perseids were supposed to peak and make jaws drop. Just like many a new observer around the country, I was disappointed not to witness any of the celestial fireworks promised by the sensationalist media reports.

Nevertheless, the episode made me look up, and I realized that on clear nights I could at least spot a few stars in the skies of Barreiro. I felt somewhat frustrated that I had not been able to figure out how to properly use a star map that came with a newspaper article as an aid to observing the shower. But I kept looking up and I started to notice some patterns, most prominently an alignment of three bright stars with two other bright stars forming a perpendicular line in the winter and spring sky.

One day I was speaking with my friend Marta Santos, a very smart and well-read girl who, like me, was a high-school student interested in heavy metal, philosophy, and the night sky. I told her about how I had been fascinated with that pattern. As soon as I described it to her, Marta, who was bound to pave a sound career as a clinical psychologist, immediately replied: "Oh, that's Orion!" And she further said: "Look, next time we get together I'll bring you this astronomy book that will help you spot and identify stars and constellations." Sure enough Marta kept her promise, and next time we met she showed up with a copy of a small book titled *Pequeno Guia do Céu* (literally, Little Guide of the Sky). That little book was my key to the wonders of the cosmos. My dad saw me so hooked on it that he soon bought me a copy, which to this day has pride of place in my personal library.

The few stars that I could see in my hometown gradually started to make sense, and one year later I was a self-fashioned expert in the Barreiro night sky (I later came to appreciate how getting started in an urban environment where only a few bright stars are visible provided me with a solid frame of reference for finding my way through darker and more densely starred skies). One day a kid in my neighborhood heard me talk about my celestial explorations and said: "Hey, if you want you can borrow a pair of binoculars that I got when I was with the Boy Scouts, I don't really use it that much." This was the piece that was missing. To this day, I will never forget the excitement I felt when I first saw the Pleiades through that 10x50 pair of binoculars! From there I would move on to co-founding a local astronomy club, working at one of the most remarkable examples of a nineteenth-century astronomical observatory (the Astronomical Observatory of Lisbon), getting a PhD in history of science with a focus on the history of astronomy, and eventually overseeing the world-class collections of the first modern planetarium of the Western hemisphere (the Adler Planetarium in Chicago). As I write, I have embarked in a new chapter of my professional life, but my fascination with the night sky remains the same. And I will always remember sitting in my office at the Adler Planetarium and thinking about how a disappointing meteor shower, a conversation with my friend Marta, a little astronomy book, and that pair of binoculars had taken me there! (Fig. 1).

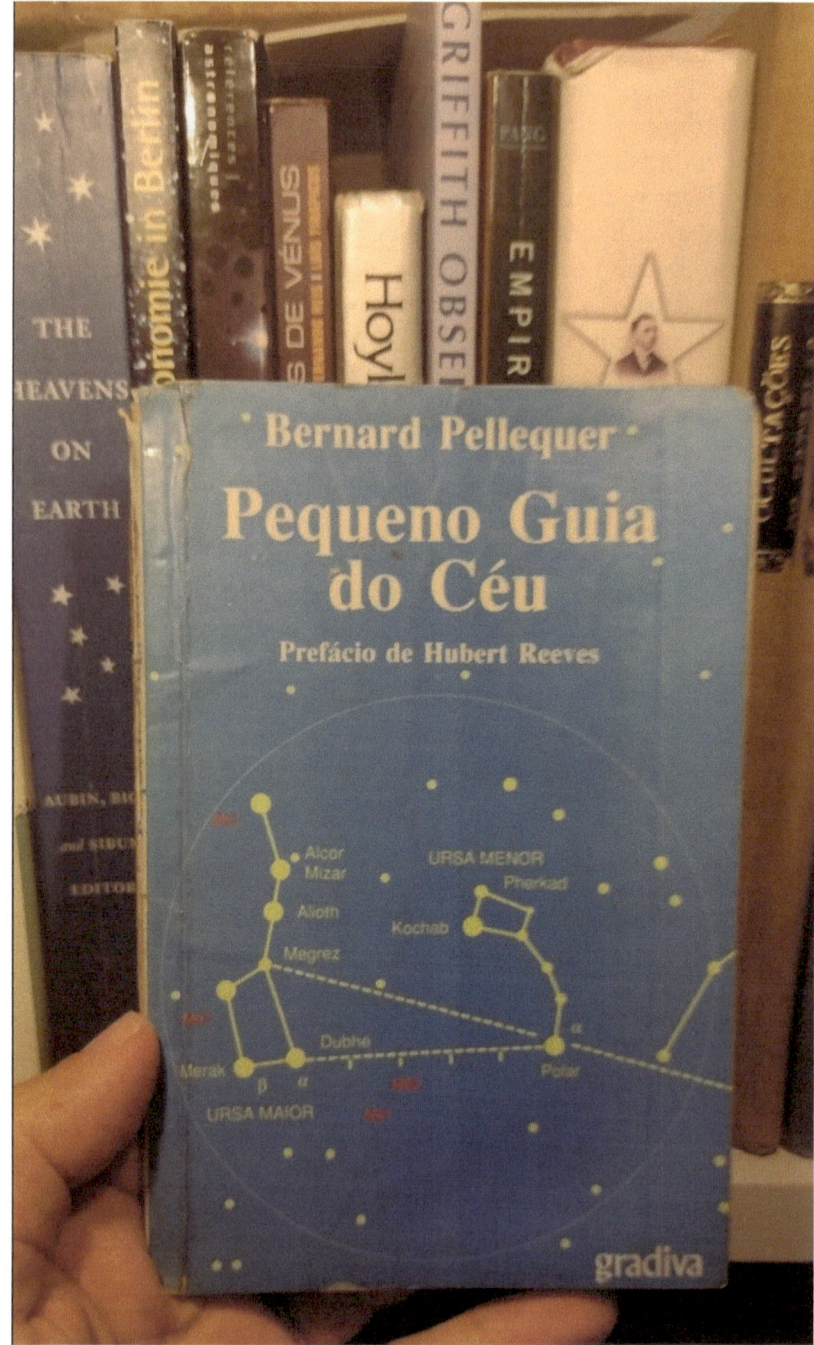

Fig. 1 My personal copy of *Pequeno Guia do Céu* (Pedro Raposo)

12   Marjorie and Roderick Webster

Pedro M. P. Raposo

The history of planetariums, especially in the United States, is filled with names of wealthy donors who sponsored the establishment of new institutions thus becoming forever associated with them—Adler (Chicago), Hayden (New York), and Griffith (Los Angeles) are just a few examples. Perhaps lesser known among the planetarium community, but still widely remembered among history of science scholars and museum curators, especially those specializing in scientific instruments, are the names of Marjorie K. Webster (1915–2011) and Roderick S. Webster (1915–1997). They did not fund the establishment of a brand-new planetarium; rather, their support was pivotal in advancing the Adler Planetarium as a major science museum and a steward of world-class collections of historical scientific instruments and rare books.

Marjorie Kelly Webster was born in 1915. While a student at Sarah Lawrence College, Marjorie spent time at Pueblo Santo Domingo in New Mexico working with Native American schoolchildren and exploring local archaeological sites. She also attended the University of California, Berkeley, and during World War II she worked for the Douglas Aircraft Company.

Marjorie and Roderick were classmates in high school. Like Marjorie, Roderick Webster was born in 1915. He was educated at the US Naval Academy and then at the University of Michigan. Roderick paved a successful career as an engineer and corporate manager, leading a foundry called Hydro Blast and working in the management of factories and mills in New England.

The former classmates married in 1953. It was nine years later that their passion for antique instruments was ignited, becoming the focal point of the remainder of their long lives.The story goes that Marjorie provided Roderick with a list of last minute supplies for a dinner party they were hosting. Roderick took the opportunity to stop by an antiques shop where he purchased a portable sundial. Eager to learn more but unable to find someone in Chicago who could educate them about their new possession, the Websters headed for the British Museum in London with the sundial. A curator there explained the piece to the couple (it was an Augsburg-type sundial made c. 1700 by Johann Martin, now in the Adler collections). He also called their attention to the Adler Planetarium in Chicago, basically in their backyard, where they would find one of the best collections of antique scientific instruments in the world. Additionally, the curator expressed his disappointment that several related inquiries he had been mailing to the Planetarium remained unanswered.

Back in Chicago, the Websters followed the recommendation and went to the Adler, where sure enough they found impressive collections of scientific instruments and rare books. They also learned about the lack of staff with the expertise to properly curate the collections, which explained the unanswered inquiries. They would soon endeavor to help change this picture.

Between 1962 and 1969, Madge and Rod, as they became affectionately known at the Adler, served as volunteer caretakers, and from 1970 until retiring in 1991, as volunteer co-curators. They also served on the board of trustees and its Collections Committee. Importantly, they developed a wide network of scholars, curators, and rare book and antique instrument dealers, which allowed them not only to develop a solid expertise in the history of astronomy and its instruments, but also to acquire a substantial number of new items for the collections, mostly with their own funds.

Even though they were essentially self-made curators and historians of instruments from a bygone era of multivalent and highly interventive benefactors, the Websters played a fundamental role in elevating the Adler's status as a museum hosting world-class collections (Fig. 2).

Fig. 2   Roderick and Marjorie Webster, date unknown (Adler Planetarium archives)

13   Katie's Story

Katie Boyce-Jacino

Before embarking on a research project that would come to define my entire academic career, I had considered planetaria with the same nostalgic fondness reserved for indoor playgrounds and roller rinks: a site of childhood pleasure, but not one I was likely to revisit. This changed during my undergraduate studies, when I studied both history and astronomy: though the disciplines began as entirely separate parts of myself, they inevitably began to merge in my mind, until I became obsessed with the history *of* astronomy. Specifically, I found myself drawn to the historical models of astronomy—the ways people of the past had endeavored to make legible the abstract world above them. I was particularly invested in orreries and armillary spheres—mechanical wonders that took the impossibly complex motions of the sky and made them accessible, tactile things.

Like many humanities graduates with little idea of what to do with myself and even less money, I embarked on that most holy of pilgrimages: graduate school. I entered my doctoral program with the vague idea that I was going to immerse myself in the gears of orreries and tellurians and armillary spheres and astronomical clocks. During that first summer, I studied at the Max Planck Institute for the History of Science in Berlin, where I read about models and spent my free time at all the cheapest cultural institutions. I found myself invariably drawn, and drawn again, to the planetarium in Prenzlauer Berg, a charming, squat late-soviet building nestled within a park. Something about the immersive illusion produced inside the planetarium felt at once familiar and entirely new—with a brain full of ideas about models and representation and astronomical pedagogy, the planetarium became a site of possibility. One evening, at a Biergarten south of the Institute, I tried to explain these feelings to my director. Two beers in, she interrupted my ramblings and said, simply: "why not write about them?" And so I did.

The planetarium has since become an enduring object of both intellectual interest and personal delight for me. As I devoted myself more and more to its history, it became clear that the planetarium could never be fully understood when considered alone, or as simply a scientific achievement. Instead, it was soon abundantly obvious that the planetarium needed to be understood entirely within the cultural context of its production—it could not be extricated from the cultural landscape of Weimar Germany in which it emerged. Once the planetarium's reach extended beyond the borders of Germany, the same framework could still apply—the planetarium demands to be understood in context. The planetarium, as this book endeavors to show, is best understood as part of a constellation of projects, past and present, to represent the sky in a theatrical, communal way. The planetarium, for me, is no longer solely an object of nostalgic fondness, but a living, evolving space that grows more complicated and more fun every time I step inside and settle down for a show.

14   I Control the Sky

Mike Smail

I'm Mike Smail, and I control the sky. For the better part of the last three decades, I've worked in four different planetariums around the United States. Worldwide, there are over 4000 planetariums, and likely tens of thousands of people who work in them. And yet, it's not a career path most people think about. It's not like the planetarium showed up at any of our high school career fairs, right? So how did I end up here?

In high school, I volunteered at a science museum in Columbus, Ohio. After volunteering for 500 hours and passing all three content training/ achievement tests, you could apply to work in the planetarium. And if accepted, you were trained on operating the equipment, presenting shows for the public, minor maintenance, all of which are pretty cool to a 15-year old kid. While I didn't have any astronomical knowledge going into it, I was able to learn on the job, and build the foundational skillsets that I'd carry with me into the future. This was one of the 1960's National Defense Education Act-funded, 9.1 meter dome, Spitz opto-mechanical star projector-equipped, planetariums that were found in hundreds of museums, schools and universities of the time. Dozens of slide projectors, the lingering smell of opaquing fluid, reel to reel audio, and enough questionable wiring in the control console to give the fire inspector heart palpitations. It was crude by modern standards, but in the mid-90 s, it was heaven.

While in college, I got a part time job at the same museum. This still involved the planetarium, but also making art with robots, training rats to play basketball, helping guests ride a bicycle across a steel rail, international video conferencing, you know, normal college job stuff. Along the way, the museum had moved into a new facility, with a new planetarium that was twice the size of the old one and featured digital projection. Instead of just projecting points of light for the stars and planets, it could display anything made of dots and lines, allowing you to show wireframe models, landscapes, and more, all in a delightful monochrome mint green. That experience using technology to create educational experiences set me up for my next planetarium at an art and science museum in Louisiana, and my current professional home, Chicago's Adler Planetarium.

In addition to seeing 500,000 museum guests a year for our self-produced, critically acclaimed sky shows, the domed theaters at the Adler play host to scientific lectures, concerts, drag shows, raves, product launches, movie/tv/music video shoots, just about any type of event you can imagine. The immersive power of space amplifies its capability; the stars were just the beginning of what a planetarium can do.

As the tools and technology evolve, the root of the experience remains. We are storytellers, using existing (and developing new) mediums to craft powerful, engaging experiences for our audiences. When I started in the field 30 years ago, the capabilities of a modern planetarium would've been considered science fiction. Trying to predict the next 30 years of planetarium technology is foolhardy. What we do know is that the planetarium dome remains a unique environment that necessitates care and a thoughtful approach to its use. All of us who fell backwards into the field are fortunate to work in it, and even more fortunate that we can develop the skills of the next generation of professional planetarians who are following us.

15   Studying the Planetarium

Matthew McMahon

When I began to work in the world of museums I did not know where I would end up. The heritage world was changing very quickly in the mid 2010's and having learned my trade in a local history museum I expected to end up somewhere similar, hopefully staying in Northern Ireland. Over the course of my Masters I worked on a World War One naval ship, now housing a museum, a coffee roastery and at the national archives. I ended up taking a short-term job at the Armagh Observatory and Planetarium while I worked on my final dissertation, planning to return to the world of traditional museums by the end of the summer.

The Board and the Director had begun to appreciate the historical importance of the institution, I was asked to begin caring for the collection and to develop policy. Over the first Lockdowns of the COVID-19 Pandemic I began to learn everything I could about astronomy, planetariums, horology and the conservation of brass. I had, like many of the children who grew up in Northern Ireland, visited the Armagh Planetarium many years before. Fifteen years later, I spent my days maintaining telescopes from the 1700's and preparing the archives for researchers when the pandemic eased. At the time I was struck by how much we knew about the Observatory, established in 1790, and by contrast how little we knew about the Planetarium which had opened to the public in 1968.

I began to rebuild the Planetarium's archive by finding the papers from management meetings and interviewing former staff to build an Oral History Archive to tell their stories. In 2022 I received an offer to study for a PhD looking at the historical geography of planetaria, through the lens of Armagh Planetarium as a case study. I did not want to stop my other work at the Observatory, designing exhibitions, supporting volunteers and caring for the archives, so I began to work on my PhD part time, and will for many years to come.

I found the mixture of theatrical, spectacular science, with the optimism and modernism of the twentieth century planetarium world incredibly interesting. The ways in which new modes of communication and travel allowed networks of professional and amateur practitioners to flourish and find each other was fascinating to track and document. I wanted to understand how these planetarium directors, some scientists, some showmen, met each other and interacted to share ideas. Ultimately, they drove each other forward to create more spectacular shows that captured the imagination of the public who had grown up in the space age. New technology, and creative inclusion of the audience are only part of it, they both stem from the desire to build experiences which inspire people to reconnect with the sky above their heads, and to experience the feeling of infinity that is kindled by a darkening dome.

16   Planetaria and Television

Nigel Henbest

It was early in 1995 when we received the call from Discovery Channel: "Hey, I want you guys to make a doco on how it would have felt to be actually on Jupiter when that comet crashed into it."

The spectacular demise of Comet Shoemaker-Levy 9 was still fresh in everyone's mind, and our TV company Pioneer Productions rose to the challenge by commissioning state-of-the-art graphics. Our documentary, *On Jupiter*, screened around the world, and garnered a Gold Medal and the Grand Award at the New York Festivals.

Even so, I always felt that a dimension was missing. On the small screen, it was impossible to evoke the visceral experience of being present in Jupiter's cloud layers. That's where a planetarium presentation would have scored. I had been a lecturer at the Caird Planetarium, Greenwich (the precursor of the

Peter Harrison Planetarium), and I knew the power of being immersed in a cosmic scene.

My experiences as planetarium presenter and TV producer taught me that a planetarium is the *sine qua non* at delivering spatial experience, while television's strength lies in telling a story that develops through time.

A TV screen is a window into an unfolding drama. That's true even of a science documentary, which—to retain its audience—must be carefully structured in a narrative arc, with a beginning, a middle and an end. *On Jupiter* succeeded because the comet impact came as the dramatic finale to a narrative that entwined the twin stories of the comet's discovery and the concurrent voyage of the Galileo Atmospheric Probe to crash into the giant planet.

A planetarium dome, in contrast, is an immersive experience. When the projector tilts to change latitude, you feel the auditorium is tilting up: even to this day, I instinctively grip the arms of my seat. A storyline is still crucial, but you can tell an effective story spatially rather than temporally. To relate the epic life of a star, for instance, pick out sequentially the Orion Nebula, the Pleiades, Sirius, Betelgeuse and finally the Crab Nebula.

There is a plethora of other media we use to communicate astronomy, of course, and in my life I've worked across most of them. It started early on when my research grant ran out, and I financed my thesis-writing at Cambridge by contributing to a part-work encyclopedia. That led to book commissions, many of which I wrote with my long-term friend and collaborator Heather Couper—together we were known as Hencoup Enterprises.

Heather was appointed Senior Lecturer at the Caird Planetarium before the siren of TV presentation lured her away. After presenting *The Planets* and *The Stars* series on British Channel 4, Heather joined ex-BBC colleague Stuart Carter and myself in setting up Pioneer Productions.

Print journalism was another rewarding way of communicating astronomy. As Astronomy Consultant for *New Scientist* magazine, I was able to fill in the stories behind the TV and newspaper headlines, including the Voyager spacecraft encounter with Neptune—faxed to London from a motel room in California—and the enigmatic 'baby picture' of the Universe taken by the Cosmic Background Explorer satellite.

Heather's *forte* was in public presentations. The format is naturally intimate, and the opportunity to ask questions involves audience members more personally—and many pick up the 'astronomy bug' by osmosis from the lecturer's enthusiasm.

Radio was an immediate, flexible and far-reaching medium. I presented a live radio commentary on BBC World Service to an audience of about 100 million when the Giotto spacecraft flew past Halley's Comet in 1986: well after the TV channels had moved on to other topics, I was among the first journalists in the world to break the news that mission control had regained contact with the apparently mortally wounded Giotto.

Television and planetaria bookend these other media. TV shows are spectacular and can wow a worldwide audience with a gripping narrative, but they are one step removed from personal experience. The planetarium experience may be smaller scale, but it's experiential, enveloping and emotional—all key qualities in enticing the audience members to fall in love with astronomy.

17   Beyond the Dome

Stephanie Ridley

Morehead Planetarium—now Morehead Planetarium and Science Center—was dedicated in 1949 as a powerful tool for the education of North Carolinians. It was the first planetarium in the world built on a college campus and only the eighth planetarium to be built in the United States. Over the past 74 years, millions of North Carolina students, teachers, and families as well as visitors from around the world have benefitted from its science programs.

True to its founding mission, Morehead Planetarium continues to bring together the unique resources of the University of North Carolina at Chapel Hill to engage the public for an enhanced understanding of science. It partners with scientists to develop outstanding planetarium shows, exhibits, classes, camps, and other programs that allow visitors to experience the wonder of science. It shows visitors why science matters and how research happening today will change lives tomorrow.

Morehead Planetarium is committed to expanding access and opportunity by delivering the Morehead experience to schools and communities across the state. Two of its signature outreach initiatives include a mobile planetarium program and the North Carolina Science Festival. Both of these initiatives exemplify Morehead Planetarium's commitment to providing high-quality science engagement opportunities for all people regardless of where they live or their ability to pay.

The Earth and Beyond program travels across North Carolina to provide enriching astronomy and earth science content to under-resourced schools and groups that might have difficulty arranging transportation to the planetarium in Chapel Hill. It utilizes a Digitalis Education Solutions portable planetarium system—a compact, high-powered projector/computer combination that projects through a fish-eye lens onto the ceiling of a 6-meter, inflatable dome. This technology allows Morehead science educators to provide a high-quality planetarium experience similar to the programs that groups receive on-site in Morehead Planetarium's fulldome theater. This mobile planetarium program was launched in 2009, and now serves more than 10,000 students each year.

Morehead Planetarium is also the proud founder and producer of the annual, month-long North Carolina Science Festival. The festival was the first statewide celebration of science in the United States, and it continues to be one of the largest events of its kind. Since its founding in 2010, the festival has reached more than 3 million participants. Morehead Planetarium engages with community partners to host events throughout the month of April in locations that include K-12 schools, public libraries, museums, parks, and science centers. These events provide attendees with an opportunity to celebrate science in fun and welcoming settings in their own communities and cultivate a positive environment that encourages children to pursue science careers.

John Motley Morehead III built a planetarium, but the science education and outreach provided by today's Morehead Planetarium extends beyond its dome and beyond its walls, opening countless minds to science every day (Fig. 3).

Fig. 3 Morehead Planetarium hosts free skywatching sessions every month (weather permitting). Morehead educators and CHAOS members (from Chapel Hill Astronomical & Observational Society) bring telescopes and guide you through fun observations of stars, planets, moons, nebulae and other celestial objects. (Jon Gardiner/UNC-Chapel Hill)

18   The Foundation of Accessible Astronomy in the Planetarium

Noreen Grice

For decades, the planetarium environment had primarily been a visual experience. As presenters, we had been taught to help visitors explore the sky visually. But one day, a group of non-visual learners came to my show. Here I share my personal journey of discovery in making astronomy more accessible using touch.

The moment that changed everything:

It was 1984. I was going into my senior year at Boston University, majoring in astronomy, and had recently been hired at the Charles Hayden Planetarium in Boston. A group of blind students were in line for the next planetarium show, and I didn't know what to do. "Just help them to their seats; that's all you have to do," is what the planetarium manager on duty told me, and that is what I did.

During the pre-recorded show, I wondered what these blind visitors thought about their experience. When the program ended, I asked them. They told me that they did not enjoy the show and walked away. To me, the planetarium was the most wonderful place in the world. This group did not share my enthusiasm, and I wanted to understand what had gone wrong.

I took a trip to the Perkins School for the Blind library in Watertown, Massachusetts. I examined their books and asked the librarian if Braille books had raised pictures. The librarian explained that raised pictures are very expensive and labor intensive, so not many Braille books have raised pictures. In that moment, I understood why the planetarium was such a poor experience: the images were projected onto the dome ceiling overhead, and the program was not pictorially descriptive. I did not know how to make astronomy more accessible, but I was determined to find solutions.

I researched organizations in the Boston area that served the blind community and found that they created tactile images by gluing string to cardboard. Each image was made individually as a custom design for the learner and indeed, it was very labor intensive.

Using etching tools and stencils, I explored making tactile images on plastic pages. I visited a teacher of the blind and his blind student to show them my work and get their feedback. They reviewed my pictures, and we had a good laugh at some of the design errors I had made. But I learned from my mistakes and created a collection of planetarium tactile images to go along with the planetarium programs.

Later, I received a grant to purchase a Braille embosser that could print multiple copies of tactile images. I redesigned all the images I had made by hand and began creating planetarium tactile books for every show. In 1990, my first published tactile book, Touch the Stars, was published. Over the years, my work has expanded to include other published books, tactile exhibits and workshops. As a consultant, I continue to build upon the foundation of tactile astronomy that I first began in 1984, reaching non-visual learners and inspiring other planetarium educators to make their facilities more accessible.

19   Enjoy! Planetarium!!

Hachioji and Mike Smail

Hachioji-san is a Japanese artist whose drawings often includes anthropomorphized depictions of planetarium projectors both historical and modern. The first ZEISS Modell I, and the Akashi Municipal Planetarium's ZEISS Universal Planetarium Projector (UPP) 23/3 that survived the Great Hanshin-Awaji Earthquake in 1995, feature prominently in their work. But you will also find stylized depictions of modern opto-mechanical projectors from GOTO, Konica-Minolta, and Megastar. Hachioji-san's art decorates keychains, pins, bags, and other types of planetarium ephemera that you might expect to find in a museum gift shop, often tagged with the label 'Enjoy! Planetarium!!'. You can view more of their artwork at https://80ooooji.booth.pm/.

The first artwork included shows the ZEISS UPP 23/3 painting the sky, with Hachioji's vivid depictions of the constellations gathered around the scene. Can you name all of them?

The following is a short text from the artist that pairs with the second included original artwork. Notice the careful eye to separate out various pieces of the Konica-Minolta Infinium Σ projector: the starball, astronomical lighting projectors (sunset, sunrise, twilight, dawn, and dusk), fulldome video projector lenses, and planet projectors, but that they all function together to produce the beautiful sky above. Toueiki is the Japanese word for projector (Fig. 4).

プラネタリウムのトウエイキさんを知っていますか?
彼はドームに星空を出現させる魔術師。
そしてオーケストラのコンサートマスター。
私たちを宇宙の広がりへ案内してくれるとても魅力的な方です。
まだお会いした事の無いならば。
次、プラネタリウムへ行った時。

Fig. 4 Some of the artwork by Hachioji (Mike Smail and Hachioji)

トウエイキさんにも会ってみてください。
意外な表情を見せてくれるかもしれませんよ?

Do you know Toueiki from the planetarium?
    He's a magician who makes a starry sky appear in the dome.
    And he's also the concertmaster of the orchestra.
    He's a very fascinating person who guides us through the vastness of the universe.
    If you haven't met him yet,
    next time you go to the planetarium,
    try to meet Toueiki.
    You might see an unexpected expression on his face!

20   Volumetric 3D Models in Planetariums

Wolfgang Steffen and Nico Koning

For several decades now, digital planetariums have made possible interactive virtual trips through the universe. One can start from Earth, fly past the planets, travel beyond the stars of the Milky Way and continue on towards thousands of galaxies at cosmic scales.

Right from the start, planetary and similar objects could be visualized quite realistically by surface mapping. More diffuse interstellar nebulae and galaxies could, however, only be represented inadequately by mapping an image on a flat or bent surface. For a more realistic visualization, volumetric 3D models were necessary. The amount of data required for volumetric models at suitable spatial resolution was not manageable by graphics hardware with the desired frame rate until about a decade ago, when the first volumetric models appeared in live interactive planetarium shows. Most of those came out of the academic computer graphics research group at the Technical University of Braunschweig, Germany, who automated 3D reconstruction algorithms from single images. This method did, however, have fundamental limitations that prevented its application to more than a few highly symmetric planetary nebulae, which are now available to view at many planetariums.

These limitations motivated us to develop volumetric 3D models based on a combination of conventional 3D modeling methods from the special effects industry and numerical astrophysics. This was the beginning of a more systematic development of volumetric 3D models for interactive visualisation as part of the digital universe in planetariums. It led to the founding of *ilumbra*, which is entirely dedicated to building volumetric models for planetariums and science centres.

ilumbra has recently developed a method which "nests" more than one volumetric 3D model inside another. This largely overcomes the key problem of the relatively low spatial resolution of these models. Key regions in a nebula or galaxy can now be represented at especially high resolution, while most of the rest of the object is at standard resolution. This reduces the memory and processing footprint enormously. With the nesting method it is now possible to interactively fly into galaxies and nebulae and maintain a crisp clear structure instead of a blurry diffuse one.

Modern long-exposure astrophotography shows more and more how many bright nebulae are embedded in a more subtle and diffuse medium that

Fig. 5 A volumetric 3D model of the Eagle Nebula (M16) with triple nesting for the "Pillars of Creation". Nesting allows interactive virtual flights into the heart of the nebula without loss of detail in the spatial resolution. (Ilumbra/RSA Cosmos)

connects the nebulae with their neighbours. With the new techniques for volumetric 3D models and the future development of graphics hardware, planetariums will be able to gradually fill our galactic environment with all the gas, dust and stars mingling continuously within. Visitors of planetariums will travel through an ever more realistic volumetric cosmic landscape (Fig. 5).

21   Future of Planetariums

Kerem Osman Çubuk

We celebrated the centenary of planetariums last year. The difference between the first planetarium systems and what is currently is simply incredible. However, most of the change has happened in the previous two decades, just like in all other tech fields. Technological advancement is still accelerating, and now we are all struggling to keep up with it.

Smartphones were not a part of our lives just 20 years ago. Today, billions of people use these devices daily for social media, banking, navigation, shopping, and leisure. Who could have guessed the current state of smartphones 20 years ago? However, considering the current rate of technological advancement, these must-have devices will probably not be part of our lives 20 years from now.

The same applies to all tech fields, including planetariums. One cannot know exactly where the planetarium world will be a few decades later. However, we can make educated guesses.

Today, one of the most time-consuming tasks for a dome show production is high-quality 3D model design and animation. The rapidly emerging AI technology will soon be integrated into 3D content production and planetarium software. Consequently, content creation will become significantly easier and more accessible.

Augmented reality (AR) will be one of the more important integrations into planetarium systems. However, it should be done without impacting the strongest aspect of planetariums: an immersive, shared experience with an unobstructed field of view.

LED dome technology, with its higher resolution, higher contrast, and seamless view, already brings incredible potential to planetariums. But this is just the beginning of a new era. Not only will dome shows become more immersive, but all activities held in LED domes, such as concerts and theatrical plays, will be like never before.

I believe the next big developments will be LED floors and frameless AR glasses in LED domes. We will get to experience the true sensation of the Ship of Imagination from Neil DeGrasse Tyson's Cosmos: A Space Odyssey. Audiences will be able to go anywhere in the universe, walk on exoplanets, dive into oceans, or experience custom-made imaginary worlds. There are no limits after this point.

Thinking of the educational potential and the new ways we could develop using these futuristic planetariums is extremely exciting! Planetariums could transform into any virtual world, and field trips could be made to planetariums for any experience. If we go one step further, each school might have a Ship of Imagination room. Maybe one day, the walls of classrooms will be covered with these LED panels, and students will wear frameless AR glasses. This would be the most significant change in classrooms since ancient times (Fig. 6).

Fig. 6 Dr Çubuk at work on making a planetarium dome show for Digistar 7 (Armagh Observatory and Planetarium)

22   The Public Legacy
of the Planetarium of Rome

Stefano Giovanardi

The new Planetarium of Rome, who took over the heritage of Rome's historical Planetarium, has been active with the current staff of astronomers and science curators for almost 20 years: from 2004 to 2014 at the Museum of Roman Civilization, under the direction of Dr. Vincenzo Vomero, it offered a variety of activities and an outstanding menu of astronomical shows, containing 65 different titles—all produced by the Planetarium staff. The shows were narrated live by the planetarium astronomers, based on a storytelling style that carefully mixes science with emotion. The appeal on the public was excellent, in fact, with a capacity of 98 seats, the visitor count was about to reach one million in ten years, when the Planetarium was suddenly closed for renovation of the Museum building.

In the following years the activity was rearranged in other locations across the city, until the original dome opened again to the public in April 2022, featuring a full dome projection system and a new original live show called "Return to the Stars". It scored over 30,000 visitors in four months: an evidence of how high were the expectations of the people of Rome about their long awaited Planetarium.

The impact of the Planetarium on society can be detected from several indicators. Some undergraduate physics students, who attended an informal school on communicating astronomy at the Planetarium, and collaborated to education and outreach events, proceeded in their careers in science working with NASA, the Italian Space Agency (ASI), the National Institute for Astrophysics (INAF), the Planetarium of Milan. Students, teachers, families and individuals often leave inspiring comments on the visitor guestbook, in their online reviews or in person after the shows. Many come back again with their friends and families.

Their feedback reveal the variety of motivations leading people to visit the Planetarium, as well as their intimate resonance with the "magic" of the experience. Here are a few examples, scattered over time: "The universe speaks / I listen, carefully / and in this sea of stars / I sink"; "Pure wonder! Not even by listening to Jimi Hendrix with the eyes closed one can reach such a level of suspension and bliss"; "After this fascinating show I felt like a star on to Earth without light"; "The embarrassment of our smallness. Thank you for letting me feel it"; "I want to become an astrophysicist. I hope the planetarium will be a good omen"; "I'll be back for the next 60 astronomical shows!"

Children, on the other hand, are in love with the Planetarium astronomer for kids, Dr. Stellarium, created by Gabriele Catanzaro to entertain the young visitors and engage them in a shared discovery of the universe. They often bring in return their drawings of celestial bodies, send in their comments and questions—to whom Dr. Stellarium replies in his podcast.

Among the ingredients of the planetarium experience, the human factor is certainly the most appreciated by the public, thanks to the quality of the live narration and of the storytelling style developed by the planetarium astronomers—a distinctive signature of the Planetarium of Rome. One recent comment says: "I had chills! Congratulations to the narrators, with their stories they guided us on an extraordinary journey. Bravo!" This is a recognition that participation and appreciation of science is greatly enhanced by the personal involvement on both sides of the scientific communication (scientists and audience).

Perhaps, the most appropriate compliment to an immersive public resource like the Planetarium, offering to everybody a comfortable space to reconnect with the universe, is expressed by the words of Sara, one of a million visitors of Rome's Planetarium: "Today I came to the Planetarium to dream".

23   Preserving the History of Planetaria

Chris Helms

As this volume has undoubtedly illustrated, planetaria—in all their various forms—have long played an integral role in not only acquainting the general public with the wonders of the heavens, but also in making the endeavor of practicing science in general more accessible and understandable. Over the centuries this has not merely been accomplished inside large dome-capped planetarium buildings or with the help of massive digital projectors. The history of planetaria is also the history of orreries, tellurians, telescopes, sundials, globes, and astrolabes. It is a history of tools and instruments, of devices and apparatus of all sorts of shapes and sizes and all manner of materials.

The Adler Planetarium contains nearly three thousand of these instruments, and similar collections exist all over the globe. The objects in these institution's care are treated and handled the same way one would care for an ancient mummy or a recently-discovered dinosaur bone. The Adler employs an entire team to ensure these historic tools and instruments are looked after and cared for in the proper manner, and to ensure they are used appropriately in exhibitions, programs, and more modern planetarium shows.

By and large, the role of a Collections Manager, or any museum collections care professional, is risk mitigation. That is, the care of these historic objects centers predominantly around preventative measures. In today's modern world it is nearly impossible to completely guarantee some unforeseen catastrophe won't befall an institution's collection, but measures can be taken to guard against the more mundane forms of artifact dangers, and to plan for various worst-case scenarios in the unlikely and unfortunate event they do occur.

From a physical standpoint, these measures include ensuring the objects are stored in the proper environmental conditions, at the proper temperature and humidity levels appropriate to their material construction. One cabinet containing stone or marble objects may require entirely different environmental conditions as the cabinet next to it containing wooden or brass artifacts. Additionally, modern lights can be extremely corrosive to many objects, especially hand-painted objects such as paper prints or gloves. Collections professionals work closely with facilities and exhibitions teams to ensure proper light exposure over time for the objects. These measures also include ensuring the objects have safe, secure, and stable enclosures such as locked waterproof cabinets to reside in when they are both on display and in long-term storage, and ensuring the objects can get the proper care if they are damaged.

Unfortunately, these measures often come at the expense of visitors having direct physical access to the objects, so part of a collections team's job in institutions such as these is to balance the need and desire of letting the public see and learn about these objects and their history, with the safety of the objects themselves.

From a policy standpoint, a collection team's job is to ensure each institution has created the appropriate policies and measures to guarantee the long-term institutional guardrails are in place to be able to care for these historic objects well into the future. These can include emergency policies and disaster plans, outlining how to act in a worst-case scenario. They can also include the creation and maintenance of more fundamental institutional policies such as those governing exactly what types of objects are collected, and who has access to them.

Caring for these historic objects is a full-time job for a collections team, and ensuring the objects are kept safe not only for future generations, but also for researchers and exhibition display can be extremely expensive for institutions. But spending the time and money to maintain these artifacts is to preserve humanity's history and to continue to carry forward the original goal of these objects—to educate and entertain people about the stars and heavens.

24   Data Visualisation on the Planetarium Dome

Arjun Chawla

It is not the strongest of the species that survive, nor the most intelligent, but the one most responsive to change. —Charles Darwin

Technological advancement in the past few decades has been exponential to put it lightly. It has vastly increased our capabilities in terms of the quality of data we need for our science as well as the quantity of data we aim to get. While we have this data, our goal not only relies in obtaining this information but analyzing it effectively for which data visualisation is without a doubt a key aspect. It is not only important for the analysis itself but to communicate our findings to our peers. Thus, the idea for visualisation on planetarium domes for astronomers needs to be analysed in several contexts which include how useful is it as a tool for our science, for communication and the feasibility of this idea.

To gauge how useful it might be as a tool for astronomy relies on the advantages it offers over a traditional computer screen. A planetarium dome offers data immersion which allows you to visualise your data and might offer new insights towards it. But ultimately, it's a tool, whether it would be useful or not depends on how one would be using it.

For communication, planetarium domes have been proven to be a brilliant method for public outreach and education. There's no reason to believe it won't be just as effective for communicating ideas at academic conferences. For day-to-day use, if the goal is immersion, a VR headset might be more practical, but on large scale communications of your results, planetarium domes just might do the trick for certain kinds of datasets.

In their 2005 article, Fluke and Bourke raised some key points for feasibility including availability and accessibility, data size and resolution, cost and lack of software tools.[11] For the availability and accessibility, their idea of investing in improving small portable/inflatable planetarium domes could be a step forward in that regard. Big planetarium domes can have a very good resolution of usually 4 K, which would help immersion. Software tools to this end are being developed by projects like data2dome[12] where they highlight as part of their mission:

- Advocating for the inclusion of dome visualization tools in standard scientific analysis and visualization packages.
- Encouraging planetaria to make their facilities available to researchers from their communities to use as a visualization tool.

Faherty et al. 2019 also propose 'IDEAS: Immersive Dome Experiences for Accelerating Science' in which they lay out a strategic plan.[13] Which highlights 2 key things, firstly, partnered software development between planetaria and academia and secondly to expand opportunities for researchers to use planetariums for data exploration.

Thus, progress is being made towards this idea as a potential tool in our arsenal for data exploration, the success of this doesn't solely depend on whether immersion would help visualisation, the crucial factor would be the participation of astronomers. Thus, to fully explore the potential of this idea, initiatives such as IDEAS and collaborations such as Data2dome need to continue, as the participation of astronomers and their visions would determine the creative boundaries we can explore.

---

[11] Fluke C and Bourke P.D. (2005), p 10–15.
[12] Science & Data Visualization Task Force - International Planetarium Society, Inc.
[13] Faherty et al. (2019).

25   Planetarium Environmental Education with Community Dialogues

Ka Chun Yu

Surveys show that although majorities of people in the United States are worried about climate change, roughly a fifth are skeptical or outright hostile to the science. Within this minority, the deficit model of science communication—the assumption that people are simply misinformed and only need more facts—is ineffective. Worse, delivering more information can backfire. Research shows that such individuals can dig into their heels even more. An alternative approach is to craft environmental change stories that highlight issues at the local level, where impacts are visible and individual action is possible. Educators can identify instances of drought, wildfire, flooding, and other extreme weather events that the audience may already be familiar with. In surveys, the general public has said they want to know more than just the basic facts; they want to learn about actions they can take to address environmental problems. A story that only invokes fear can lead to an irrational, emotional response, while a story that triggers a mix of hope and worry can motivate people to think about their choices and act. To convey hope, narratives can showcase projects where groups are making a difference in their own small ways to repair environmental damage. Such examples show that these problems are not intractable, and that many groups and individuals are creating plans to deal with the impacts from climate and other environmental change.

The best practices just described gave inspiration to the creation of the Worldviews Network (www.worldviews.net). This collaborative web of interdisciplinary scientists, artists, and educators was funded by NOAA from 2010–2014, and turned the planetarium, a space traditionally used for astronomy education, into a venue for facilitating dialogues around local issues at each partner institution. The Network created more than a dozen "bioregional community dialogues." Audiences viewed dome presentations with realistic flights over a virtual Earth, designed to replicate the "Overview Effect," the perspective shift that astronaut returning from space reported having that made them more aware of the fragility of Earth's biosphere and the need to protect it. The stories were created with the collaboration of experts, including those from local organizations that not only had knowledge about a topic, but also had on-the-ground expertise in devising solutions or mitigating impacts.

The presentations were built around the actions of see, know, and do. "Seeing" involved experiencing immersive storytelling augmented with

Fig. 7 Using a digital planetarium to engage audiences about Earth systems at the Denver Museum of Nature & Science (left), and a post-dome discussion session at the California Academy of Sciences (right). (Denver Museum of Nature & Science and Ka Chun Yu)

geospatial datasets. "Knowing" meant a deeper, cross-disciplinary understanding of Earth's interconnected systems. "Doing" was via interaction among the participants (including the invited outside experts) to help audience members remain engaged in the problems identified in the story. The Worldviews Network helped create a diverse set of dialogue topics at a wide variety of institutions, from large natural history museums to science centers to small non-profit educational organizations. These events allowed participants to understand the complex geophysical, biological, and human systems that interact with each other; helped audiences visualize phenomena at different scales from the global to the local; and allowed them to envision themselves as proactive participants in helping to affect change in their communities (Fig. 7).

Acknowledgement   This entry is a shortened version of Nazé **2023** (JAHH, vol **26#4, 816**). The author acknowledges the use of AFOEV, ADS, gallica, Roubaix mediathèque, archive.org, and jstor databases.

References

Bergeron A, Bigg C (2015) D'ombres et de lumières. L'exposition de 1937 et les premières années du Palais de la découverte au prisme du transnational. Revue Germanique Internationale 21:187–206. https://doi.org/10.4000/rgi.1529
Chilton L (2017) 1947–1971: The L.A.A.S. Griffith years of telescope making. The Bulletin of the Los Angeles Astronomical Society 91(5)

Faherty JK, SubbaRao M, Wyatt R, Ynnerman A, Tyson ND, Geller A, Emmart C (2019) Ideas: immersive dome experiences for accelerating science. https://doi.org/10.48550/arXiv.1907.05383

Nazé Y (2023) Reysa Bernson, the unconventional head of the first French planetarium. J Astron Hist Heritage 26(4). Astral Press, Floreat, Australia

King DA (2007) Henry C. King (1915–2005). J Hist Astron 38(4). SAGE Publications Ltd:526–527

King HC, Millburn JR (1978) Geared to the stars: the evolution of planetariums, orreries, and astronomical clocks. University of Toronto Press, Toronto

King HC (1979) The history of the telescope, 2nd edn. Dover Publications, New York

Silva EC (1935–1942). (unpublished observing log, library of the Calouste Gulbenkian Planetarium)

Silva EC (1948) Astrophotography Atlas, Alfeite (unpublished photographic albums, library of the Calouste Gulbenkian Planetarium)

Silva EC (1958) Atlas Astrofotográfico, (unpublished photographic albums, library of the Calouste Gulbenkian Planetarium)

American Institute of Physics (2015b). https://www.aip.org/history-programs/niels-bohr-library/oral-histories/32485

Fluke CJ, Bourke PD (2005) Digital Planetariums for everyone: Astronomy visualisation in reflection. The Planetarian

Williams T (2000) Getting organized: a history of amateur astronomy in the United States. PhD thesis. Rice University